IMAGES
of America

TANNEHILL IRONWORKS

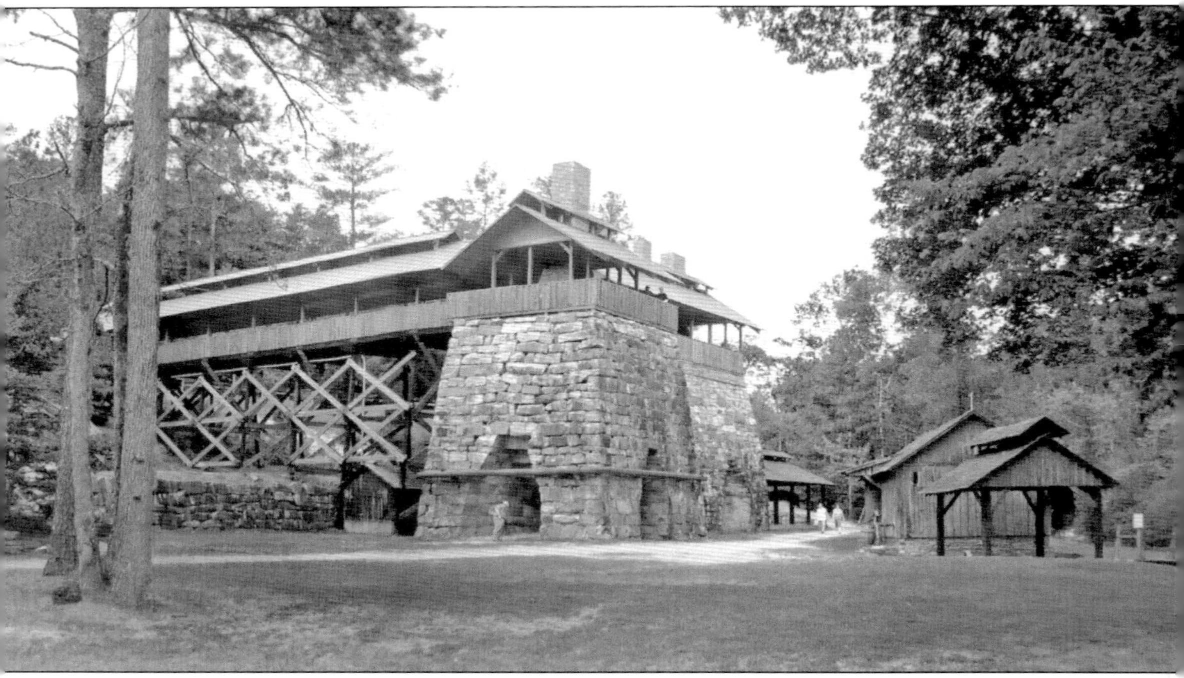

The Tannehill Furnaces, near Birmingham, Alabama, rank among the nation's best-preserved ironworks from the Civil War era. As the war drew to a close, they were badly damaged by Union troops on March 31, 1865. Today, the restored iron maker is the centerpiece of Tannehill Ironworks Historical State Park, a 2,063-acre industrial heritage site in Jefferson, Tuscaloosa, and Bibb Counties. (Marshall Goggins.)

ON THE COVER: This photograph, taken by Birmingham photographer John T. Horgan of an American Institute of Mining Engineers outing in May 1888, shows early extraction work at the Tannehill Iron Ore Mines. Ore mining would expand to cover the area from Bucksville to Woodstock over the next decade. (Alabama Department of Archives & History.)

IMAGES
of America

TANNEHILL IRONWORKS

James R. Bennett

ARCADIA
PUBLISHING

ISBN 978-0-7385-8241-2

Published by Arcadia Publishing
Charleston, South Carolina

Printed in the United States of America

Library of Congress Control Number: 2010940835

For all general information, please contact Arcadia Publishing:
Telephone 843-853-2070
Fax 843-853-0044
E-mail sales@arcadiapublishing.com
For customer service and orders:
Toll-Free 1-888-313-2665

Visit us on the Internet at www.arcadiapublishing.com

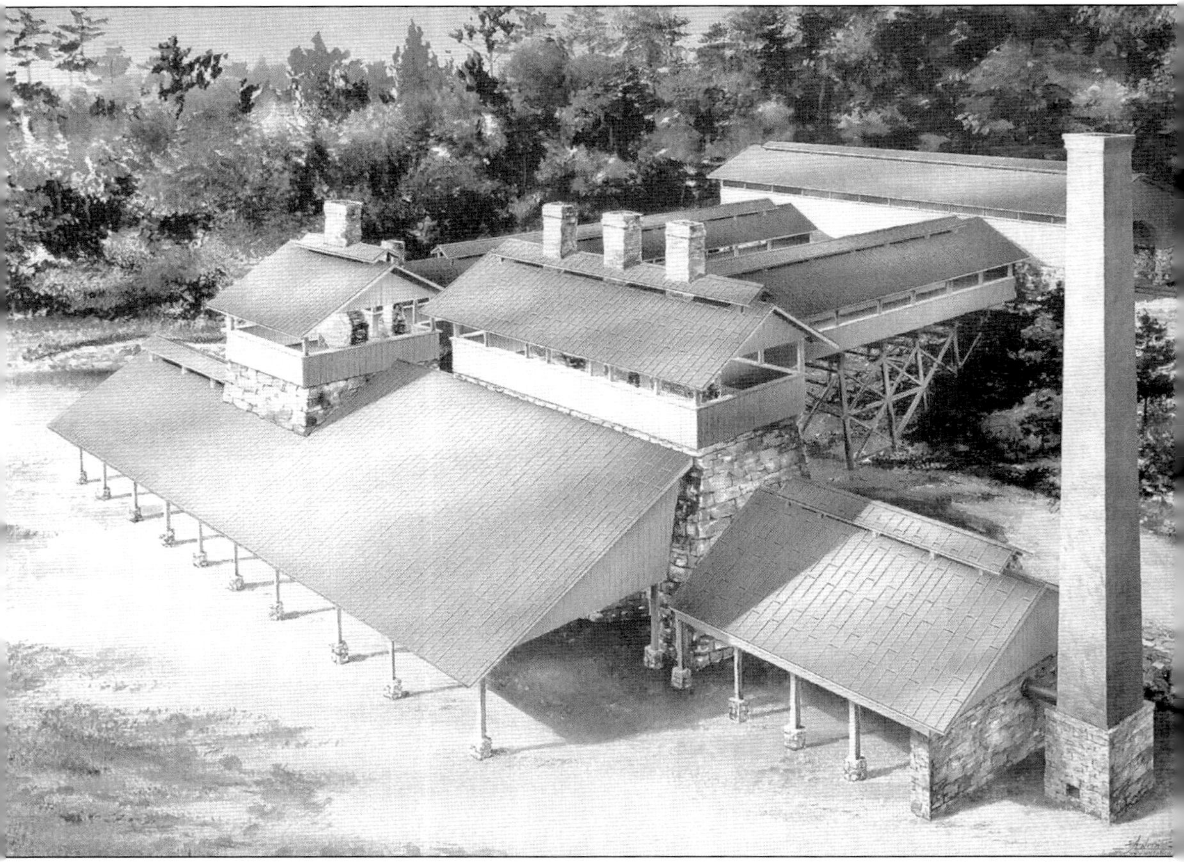

This conceptual illustration of how the Tannehill Ironworks appeared during the Civil War is part of the restoration plan for the furnace complex. The drawing was done by Anita Bice for architect Vic Fortinberry and is based on historical and archaeological evidence. (Anita Bice.)

Contents

ACKNOWLEDGMENTS

"The only thing new in this world," Harry Truman once wrote, "is the history you do not know."

To be certain, there are many lost stories tucked away in the ore-rich hills around Old Tannehill where the Birmingham iron district got its start. From 1830 through the close of the Civil War, hundreds of miners, furnace workers, and teamsters literally made history each day of their extended labors. In this work, we attempt to shed some light on a few of their experiences through photographs assembled over the years from dozens of collections.

Appreciation is extended to the staffs of the Alabama Department of Archives & History, the Birmingham Public Library Archives, the special collections libraries at both Samford University and the University of Alabama, the U.S. Geological Survey and Photographic Library, and the Iron & Steel Museum of Alabama.

Special thanks are expressed to Elizabeth Wells, Rickie Bruner, Marty Everse, Jim Baggett, John Norris, Lewis Dean, Nick Tew, and Ralph Pate.

Unless otherwise noted, all images appearing in this book were graciously provided by the Iron & Steel Museum of Alabama at Tannehill.

This book is dedicated to the hard work of the employees of Tannehill Ironworks Historical State Park from 1970 to the present.

Introduction

Having had its start as a bloomery along the banks of Roupes Creek in 1830, the Tannehill Ironworks, located near the Jefferson and Tuscaloosa County line, grew into one of the most productive centers of iron manufacture in Alabama during the Civil War.

The only three-furnace ironworks in the state during the war years, this important battery had a rated daily capacity of 22 tons, almost all of it destined for the CS Naval Gun Works at Selma.

Shipped along the rails of the Alabama & Tennessee River Railroad from Montevallo, Tannehill iron was cast into shot and shell, gunboat plating, machine parts, and all the implements of war.

Equipped with hot blast stoves and steam power to increase production, this trio of tall stone furnaces was among the most modern of their day. Experiments with red iron ore in 1862 opened the door to iron manufacture later in Birmingham.

A foundry and tannery were operated in connection with the ironworks, producing cast iron and leather products, like kettles, ovens, hollowware, harness, and canteens, for the Confederate army. There is evidence that the foundry at various times also made bells, machine parts, nails, and custom castings.

Furnace No. 1 was built by noted Southern ironmaster Moses Stroup in 1859. He also began work on Furnaces Nos. 2 and 3. William L. Sanders, a merchant from Selma, completed the expanded plant, which went into blast in 1862–1863 with financial help from the Confederate government when Stroup left to build the Oxmoor Furnaces in Jefferson County.

Hillman Bloomery, erected when Andrew Jackson was in the White House, is itself an important industrial heritage site. The work of Daniel Hillman, who moved here from Kentucky in search of an iron fortune, the dual-chambered Catalan forge made horseshoes, plows, and other agricultural products for markets in Jonesboro, Elyton, and Tuscaloosa. Its location near Tannehill Furnace No. 1 is considered the birthplace of the Birmingham iron and steel district.

No other place in Alabama made iron for as long as the Tannehill site prior to the Civil War.

Alabama furnaces—including those at Tannehill—accounted for 70 percent of the Confederate iron supply in the last two years of the Civil War, considering Federal raids had destroyed ironworks in other Southern states.

To hasten the end of the hostilities, Union war planners, including Generals Grant and Thomas, sent 13,480 Federal soldiers into Alabama during the winter of 1864 to destroy the iron industry and the huge foundries and arsenal at Selma. This they accomplished in March and April 1865 in what would be the largest cavalry campaign of the war.

Were it not for the assignation of Abraham Lincoln on April 14, this story would have loomed even larger in American history.

Armed with new seven-shot Spencer repeater carbines, the Federal force under Gen. James H. Wilson culminated the Alabama campaign by overwhelming Confederate general Nathan Bedford Forrest's cavalry at Selma on April 2. Twelve Alabama iron producing companies fell to union torches during the raid and other Federal incursions. Only the works of Hale & Murdock in Lamar County, which is located off the line of Wilson's march, survived.

Under command of Capt. William A. Sutherland, three companies of the 8th Iowa Cavalry attacked the Tannehill Furnaces, also known as the Roupes Valley Ironworks, on March 31. At day's end, the furnaces were no longer operational, and the foundry, tannery, gristmill, and a tax-in-kind warehouse were in ruins. Also set to the torch was residential housing used by both white workers and slaves.

George Monlux, a 21-year-old commissary sergeant in Company I, wrote in a report that the furnace workers must have been forewarned of the impending attack since very few workers were present when Sutherland arrived. However, shots may have been fired at some of the workers fleeing the mill, which would account for a spent Spencer cartridge found during an archaeological excavation of the slave quarters in 2009. Only the Federal force was equipped with Spencer carbines, a major upgrade in firepower.

Although the official records indicated three companies of Federal cavalry were present during the furnace raid, Monlux lists only two; the third company was probably dispatched in search of the Williams & Owen Forge, one mile downstream.

Sutherland's detachment rejoined the column led by Brig. Gen. John T. Croxton, the brigade commander, and headed toward Tuscaloosa to burn the University of Alabama, which was thought to be a military school. There they also torched the factories in the town, including the Leach & Avery Foundry, a cannon maker.

Sutherland served as Croxton's adjutant general. At Tannehill, his soldiers made quick work of the place fearing General Forrest had troops in the area. Before the day was done, the Federal raiders had damaged all three furnaces, wrecked the steam engine, set fire to the overhead charging bridges, tore up the tramway to the ore mines, and torched the foundry and cast houses. Fire was also set to the blower house, the location of a large waterwheel that supplied air blast to Furnace No. 1. In his report, Monlux wrote the following:

> These works were of great size and must have covered more than an acre of ground and near the [furnace] building were tens of thousands of bushels of charcoal. The boys started fires in several places in the building [and] as it was built of pitch pine the flames spread so rapidly that some of the boys came near being burned before they could get out. We watched the building and these large piles of charcoal until they were well on the way to destruction when we started on to intercept the column.

The only people the Federal troops encountered were a group of women slaves. One of these women told Monlux she was "powerful glad" the Union army was putting the works out of commission because it "used up" many of her people every year. Industrial injuries and even fatalities were often recorded at blast furnaces.

Archaeologist Jack Bergstreser suggests the lady with the nerve to speak so openly may have been the chief cook at the nearby slave quarters, where excavations have revealed a cooking area serving a group of more than 60 workers. Today, the stone chimney bottoms in the worker housing area bear mute evidence to the past lives of this long-forgotten industrial workforce. Slave beads, ceramic smoking pipes, and pieces of broken cookware and pottery mark the old homesites.

The furnaces at Tannehill never again operated on a commercial scale, although scrap was used in the plant's cupola several years after the war. A cupola is a small furnace designed to remelt furnace iron and increase its carbon content for castings by use of coke as its fuel source. From 1865 to 1867, both B.J. Jordan and James T. Loveless operated the cupola using scrap and pieces of pig iron for small castings.

While iron production ceased in 1865, Furnace No. 1 was repaired and put back into operation for a run of iron during the American Bicentennial Celebration in 1976. The Smithsonian Institution said it marked the first time in American history that an iron furnace out of blast for over a century had ever been refired.

The event, witnessed by more than 15,000 people on September 19, attracted the participation of dozens of Birmingham iron and steel companies. Some 2.5 tons of iron recovered from the operation was cast into souvenir ingots and several full-sized cannons.

Several attempts were made after the Civil War to rebuild the Tannehill plant, including one by a company in Pennsylvania, but these efforts all fell through. The mining of brown ore near

Tannehill, however, was resumed by the Pioneer Mining & Manufacturing Company, a predecessor to Republic Steel Corporation, in 1887.

An article appearing in the *Scientific American* on March 27, 1869, revealed that the Pioneer Company, headed by David Thomas, had acquired the ore fields the previous year along with the ironworks and intended to invest $500,000 in them.

Instead, Giles Edwards, a former assistant superintendent at the Shelby Iron Company who was engaged as a land agent for the Pioneer Company, built a new plant eight miles to the west at Woodstock in 1880, the first Alabama blast furnace blown in on coke. He mined ore a short distance from his furnace.

Edwards, a native of Wales who came to America to work in the Pennsylvania iron industry in 1842, moved into the old Mansion House at Tannehill in 1871. He is thought to have used salvageable equipment from the burned-out plant in his new furnace. In doing so, he may have become the original scrap collector at Tannehill, which was heavily pillaged years later during both world wars.

His plucky wife, Salinah, joined him in building the Woodstock plant from 1873 to 1880. Since they had little help, it took them seven years to complete the construction.

"On our way to Tannehill we passed through Elyton and saw the site of Birmingham," she wrote. "There were then only two section houses for the men starting the railroad, nothing else but my husband pointed up the long valley. There lies Birmingham, he said . . . all that is going to be Birmingham some day and he spread his arms out to take in the whole country," said Salinah.

Birmingham was founded in 1871—six years after the war had ended.

In gearing up for the bicentennial event, the state spent over $100,000 in making repairs to Tannehill Furnace No. 1.

The restored ironworks today is listed on the National Register of Historic Places and the international landmark registry of the American Society for Metals. The site has also been placed on the Civil War Trust's National Discovery Trail and included in the Alabama Engineering Hall of Fame.

The old furnaces are the centerpiece of Tannehill Ironworks Historical State Park, located at the corners of Jefferson, Tuscaloosa, and Bibb Counties and southwest of Birmingham. Administered by the Alabama Historic Ironworks Commission, a state agency, the park annually attracts over 400,000 visitors. It is the most visited Civil War site in the state.

In addition to the ironworks, the park is also the location of 45 historical buildings, many of them dating to the 1800s; the Iron & Steel Museum of Alabama; the John Wesley Hall Gristmill; and the May Plantation Cotton Gin House.

The park has become a national model on how to transform a former industrial site, out of use for over a century, into a major tourist attraction.

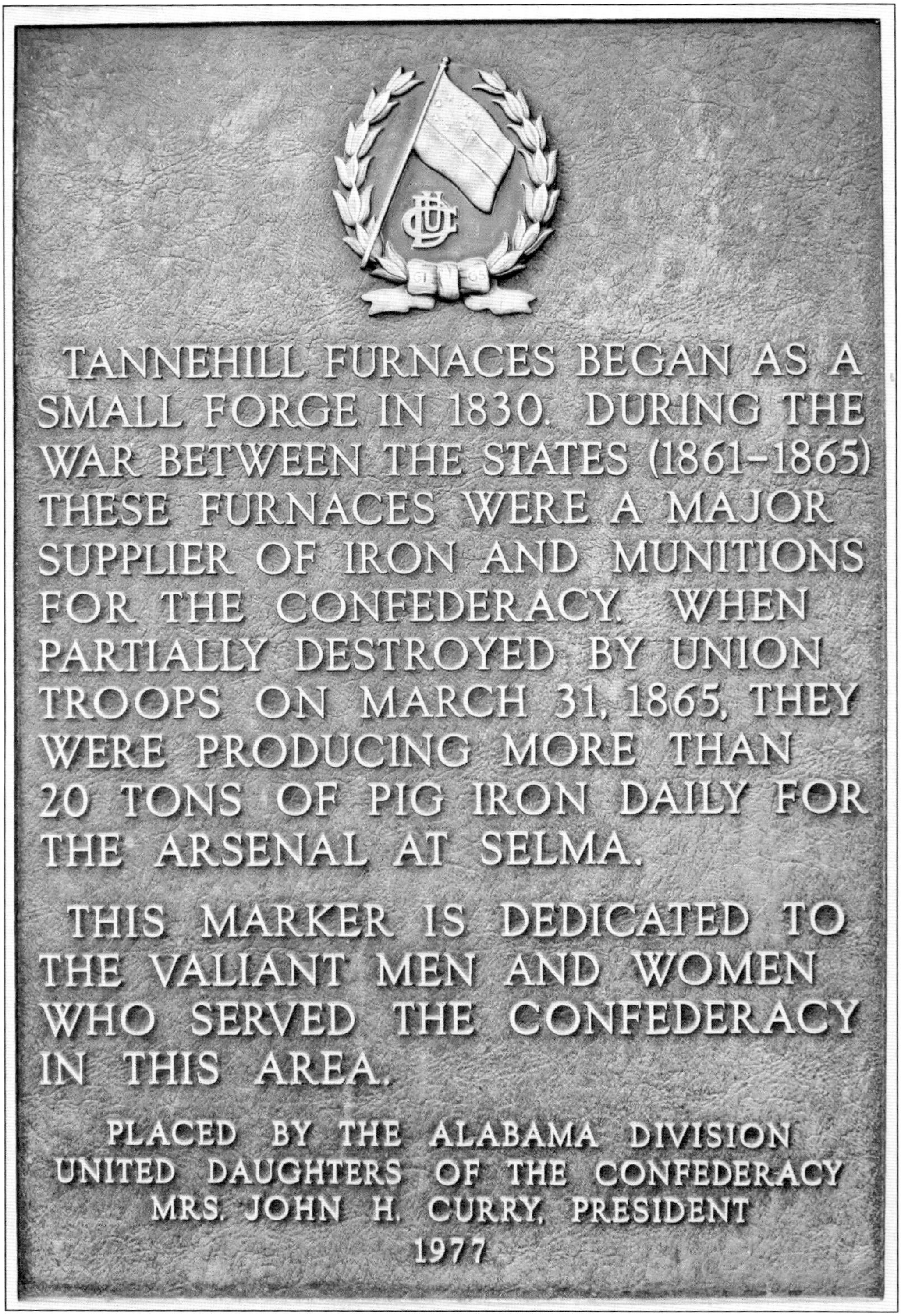
TANNEHILL FURNACES BEGAN AS A SMALL FORGE IN 1830. DURING THE WAR BETWEEN THE STATES (1861–1865) THESE FURNACES WERE A MAJOR SUPPLIER OF IRON AND MUNITIONS FOR THE CONFEDERACY. WHEN PARTIALLY DESTROYED BY UNION TROOPS ON MARCH 31, 1865, THEY WERE PRODUCING MORE THAN 20 TONS OF PIG IRON DAILY FOR THE ARSENAL AT SELMA.
THIS MARKER IS DEDICATED TO THE VALIANT MEN AND WOMEN WHO SERVED THE CONFEDERACY IN THIS AREA.
PLACED BY THE ALABAMA DIVISION
UNITED DAUGHTERS OF THE CONFEDERACY
MRS. JOHN H. CURRY, PRESIDENT
1977

One

Hillman Bloomery

While Catalan forges and their later configuration, the American bloomery, were the precursors of high stone blast furnaces, they continued to coexist until after the Civil War. By 1856, there were 17 Catalan forges and bloomeries known to exist in Alabama. Two of them would be built at Tannehill.

Daniel Hillman, who honed his iron-making skills around Philadelphia as a youth, eventually took his metal-working skills to Ohio, Kentucky, and finally, Alabama, where, with financial help from Abner McGehee, he put up a dual-chambered bloomery on Roupes Creek in 1830. This site is considered the birthplace of the Birmingham iron industry.

The Hillman Bloomery produced iron that sold for 10¢ a pound. The operation had a capacity of 300 pounds a day and commanded a market from Tuscaloosa to Elyton, site of present-day Birmingham. It would operate until replaced by the larger Tannehill Furnaces, also known as the Roupes Valley Ironworks, in the late 1850s.

A second bloomery, the Williams & Owen Forge, was built a mile downsteam in 1861 to provide iron for the local market, but soon, its product also found its way into military uses during the Civil War.

Hillman died unexpectedly in 1832 and is buried in an unmarked grave in the Bucksville Baptist Cemetery. His son and grandson would become famous in their own right in the iron business.

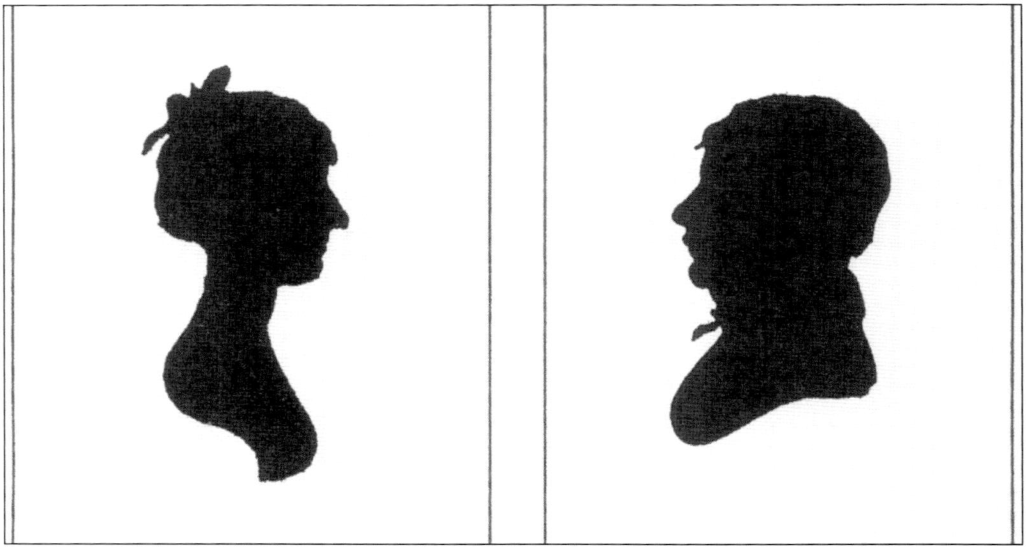

These silhouettes of Daniel Hillman Sr. (1782–1832) and his wife, Grace Huston (1773–1826), are taken from the Hillman family Bible around 1806. After his wife died, Hillman came to Alabama to build the forge on Roupes Creek. Hillman built forges in New Jersey, Ohio, Kentucky, and Alabama. (Charles Priest Jr., McLean, Virginia.)

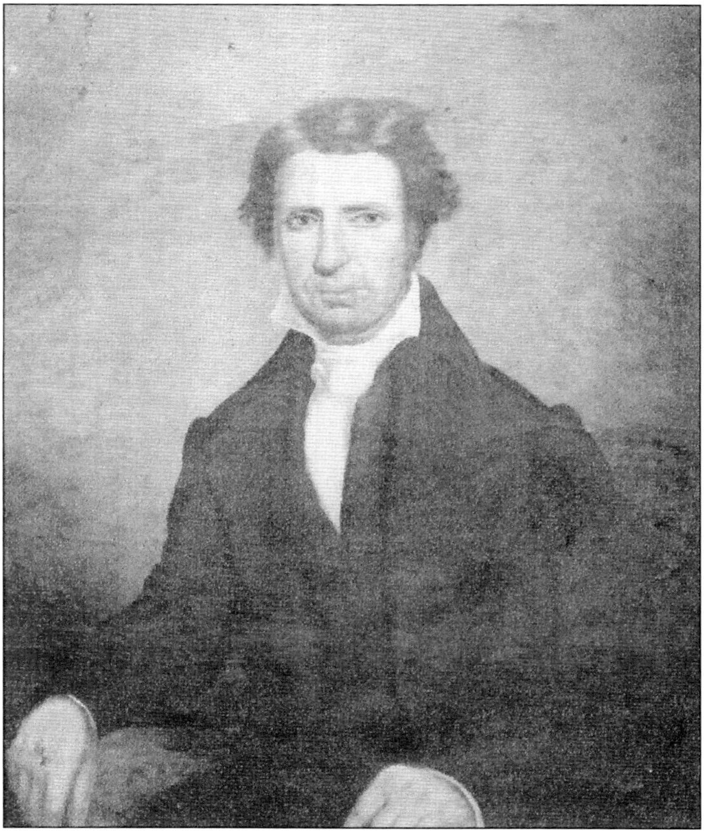

Abner McGehee, a wealthy cotton planter from Hope Hull in Montgomery County, was the major investor in the Hillman Bloomery. He acquired the Tannehill forge track, which was about 156 acres in Tuscaloosa County, through two land grants dated 1819. McGehee invested heavily in railroads, plank roads, and hotels and built the first railway streetcar system in Montgomery, Alabama. (Alabama Department of Archives & History.)

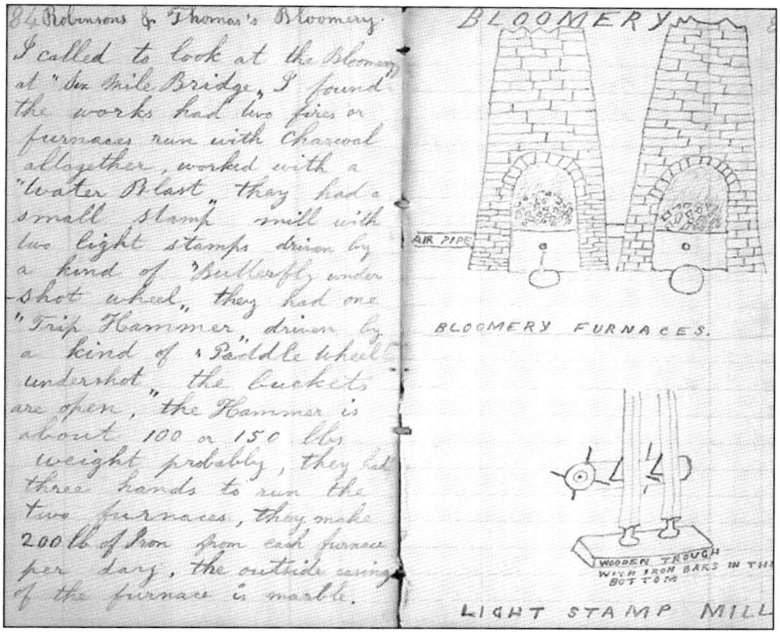

84 Robinsons & Thomas's Bloomery.
I called to look at the Bloomery at "Six Mile Bridge," I found the works had two fires or furnaces run with Charcoal altogether, worked with a "Water Blast" they had a small Stamp mill with two light stamps driven by a kind of "Butterfly under shot wheel," they had one "Trip Hammer" driven by a kind of "Paddle wheel undershot" the buckets are open," the Hammer is about 100 or 150 lbs weight probably, they had three hands to run the two furnaces, they make 200 lb of Iron from each furnace per day, the outside casing of the furnace is marble.

BLOOMERY
AIR PIPE
BLOOMERY FURNACES.
WOODEN TROUGH WITH IRON BARS IN THE BOTTOM
LIGHT STAMP MILL

Here is a drawing of Bibb County's Six-Mile Bloomery from Joseph Squire's Journal in 1868. Also known as Robinson & Thomas's Bloomery, the dual-chambered iron forge had a waterwheel for air blast and a small stamp mill for crushing ore. Erected in 1863, it was located about 25 miles from Tannehill and could produce 400 pounds of iron daily. (Ken Penhale, Helena, Alabama.)

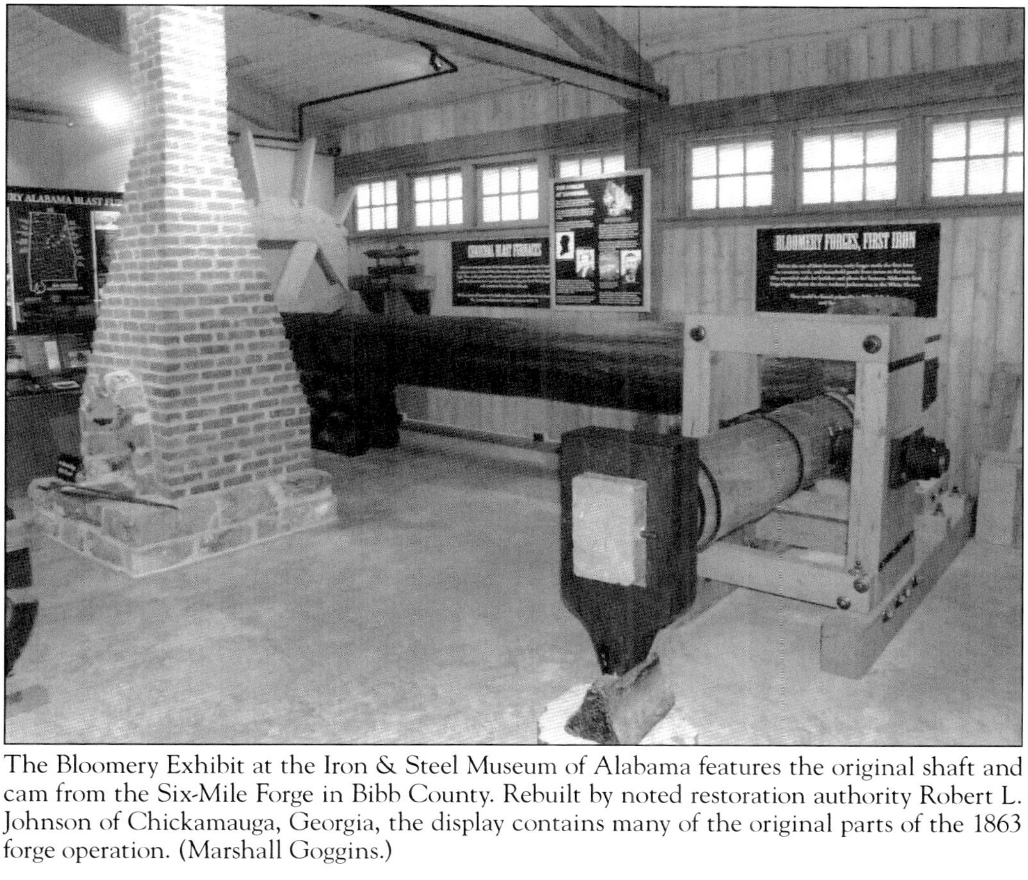

The Bloomery Exhibit at the Iron & Steel Museum of Alabama features the original shaft and cam from the Six-Mile Forge in Bibb County. Rebuilt by noted restoration authority Robert L. Johnson of Chickamauga, Georgia, the display contains many of the original parts of the 1863 forge operation. (Marshall Goggins.)

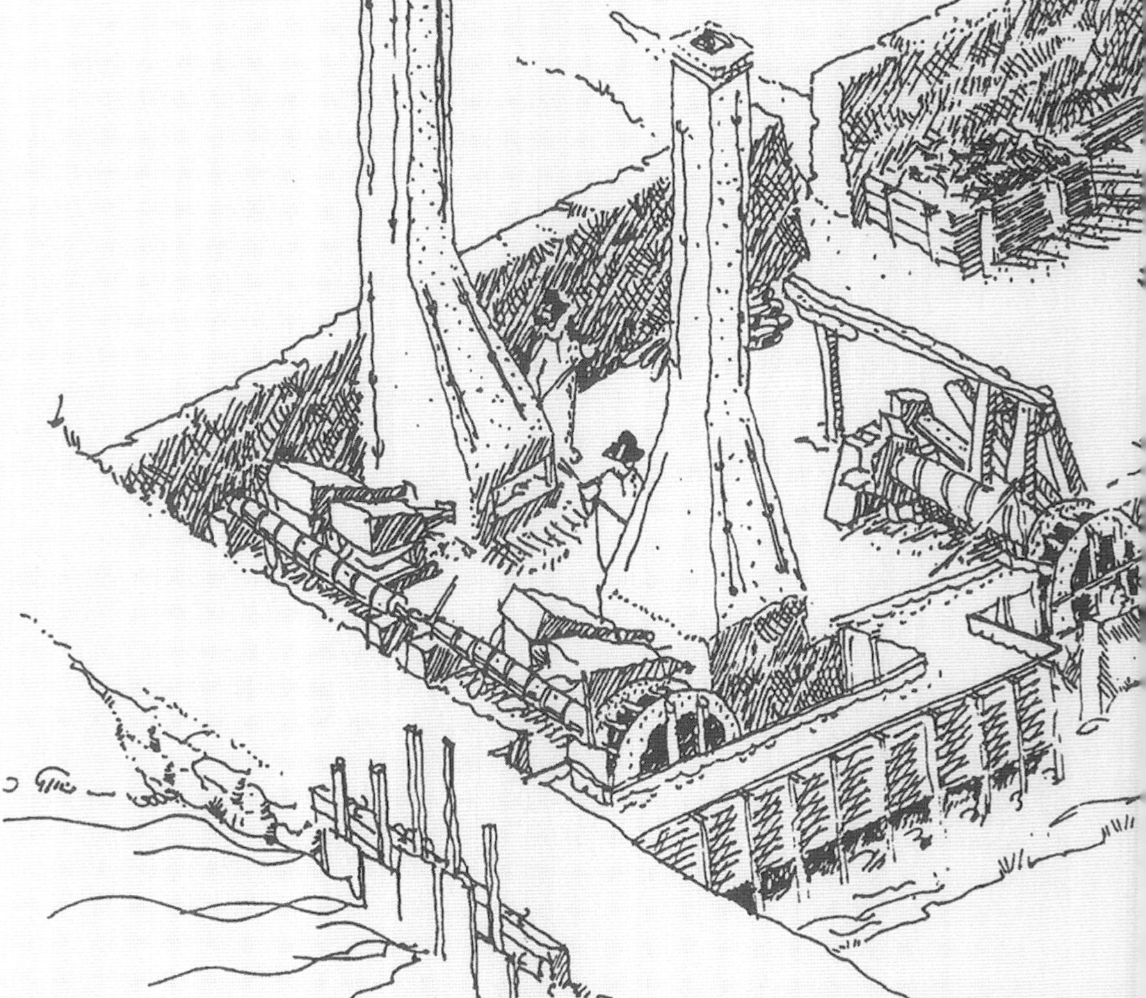

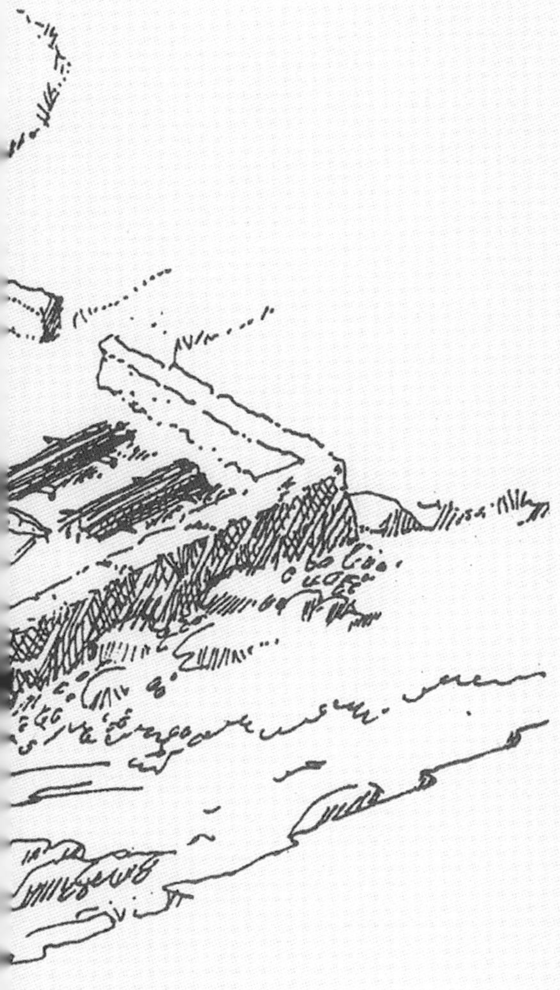

IERY FORGE

This Swedish bloomery forge is thought to be much like Hillman's bloomery at Tannehill with two firebox chambers. It shows the forge hammer and its waterwheel, as well as the waterwheel powering both of the leather air bellows at the rear of each forge hearth. The cams on the waterwheel shafts lifted the hammer and closed the air bellows. Iron produced at bloomeries, which was called bar iron, had low carbon content and therefore was quite workable. It was considered good iron for blacksmiths, as it could be bent cold without breaking. Alabama had 17 known Catalan forges by 1856. Samuel Wear built the first forge of record around 1825 on Shoal Creek in Shelby County. (Karl-Gustaf Hildebrand, 1992, *Swedish Iron in the Seventeenth and Eighteenth Centuries, Export Industry Before the Industrialization*.)

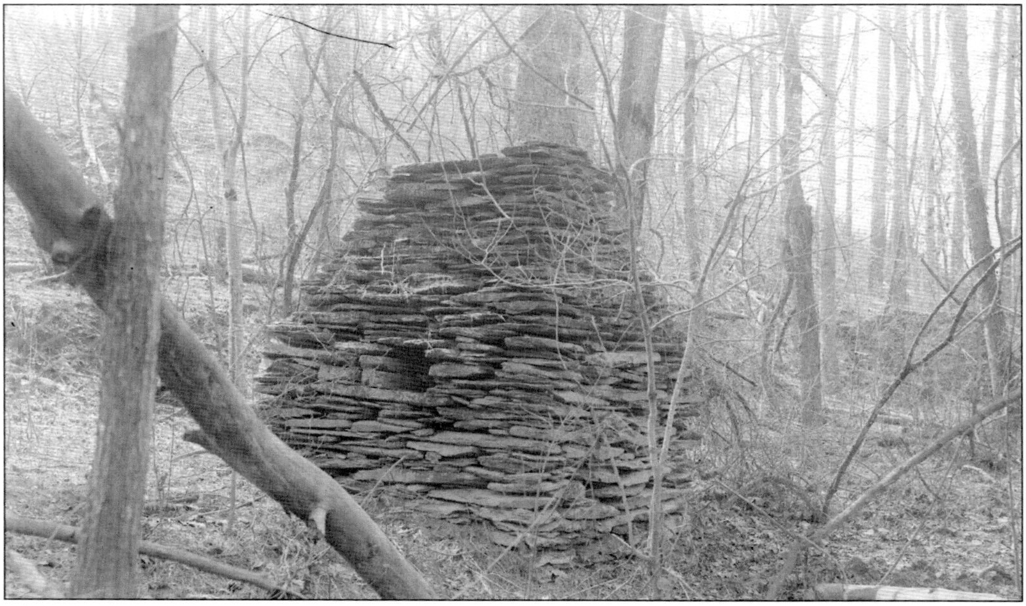

This photograph shows the remains of Eagle Forge around 1910 on Talladega Creek, near Waldo. Its rear section, where the blast equipment was connected, is pictured. Erected in 1846 by George D. Wheeler and Israel Sprayberry, it was located three miles from Maria Forge. Built next to creeks, many of these structures did not survive floods, and few remnants remain today. (W.S. Hoole Special Collections Library, University of Alabama.)

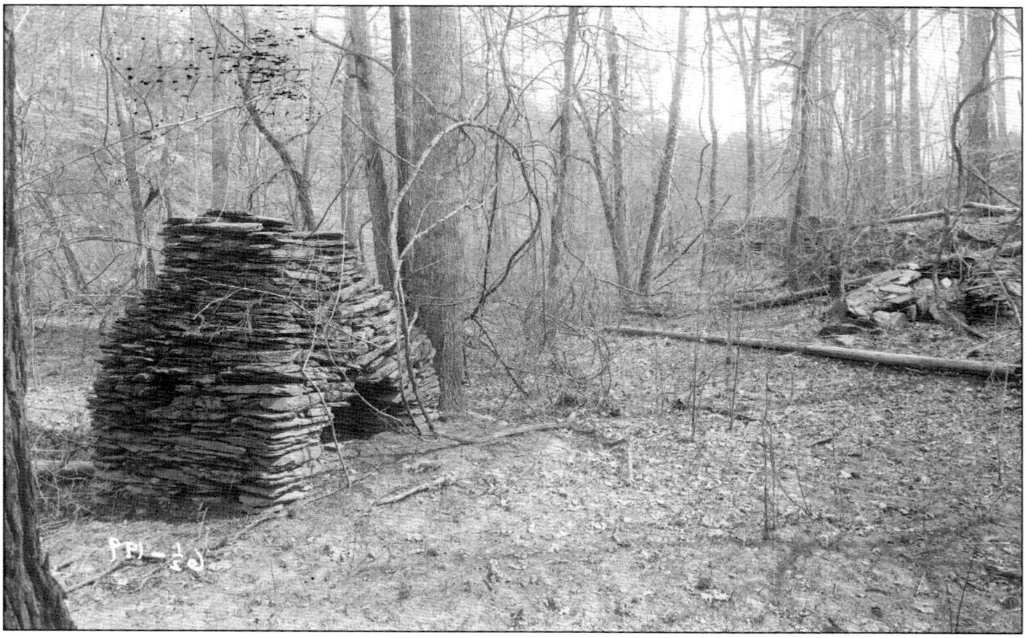

This is the front view of the old Eagle Forge and its firebox. This Catalan forge was somewhat smaller than the nearby Maria Forge but made bar iron for local smiths. Floodwaters badly damaged it; however, Eagle Forge was one of the last to survive. (W.S. Hoole Special Collections Library, University of Alabama.)

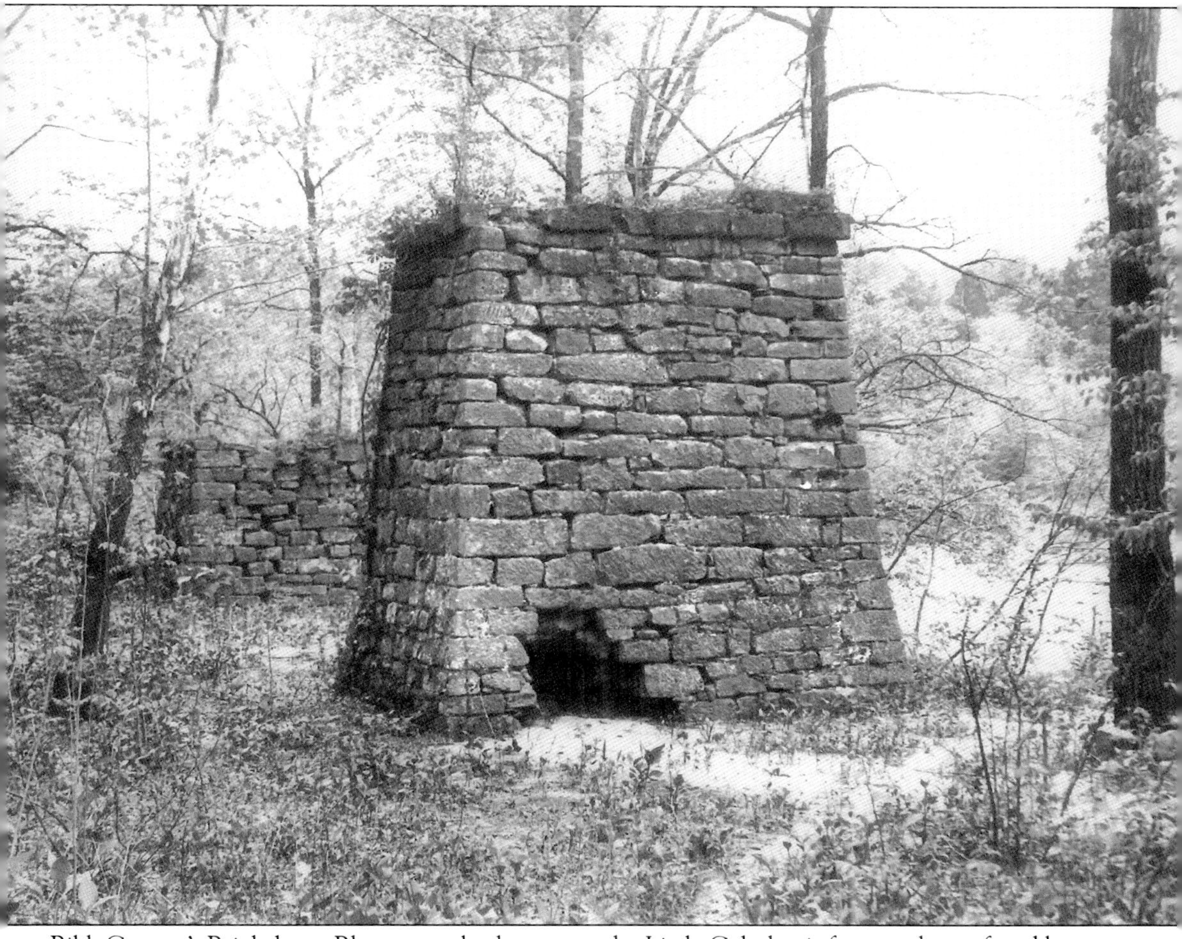

Bibb County's Brighthope Bloomery, also known as the Little Cahaba, is frequently confused by some historians with a blast furnace, although slag found at the site indicates it was a forge. Shown here in a ruinous condition during the 1930s, it was built by William P. Brown with the help of local ironmaster Jonathan Newton Smith in 1847–1848. A second and larger plant was added in 1863 but damaged during the Civil War. Brown, a lawyer and legislator, was one of the original owners of the Montevallo Coal Mines, which he continued to operate until his death in 1868. He was under contract to provide coal to the Selma Arsenal. With the help of the Confederate government, the W. L. Ward Company leased and expanded the Brighthope Bloomery, but it may have had an intermittent operation. Ore used in the ironworks was mined two miles away. (Alabama Writers' Photograph Collection, Alabama Department of Archives & History.)

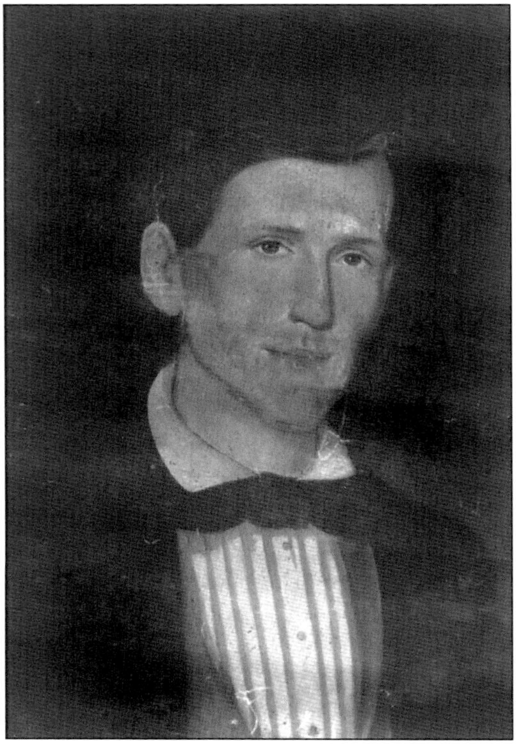

At 22 years old, Marion Tannehill, the son of the second owner of Hillman's Forge, Col. Ninian Tannehill, was listed as ironmaster at the plant in the 1850 census. He married Sarah Elizabeth McMath, the daughter of a wealthy local planter, on October 15, 1846. Marion later became a physician. Sarah's father, Elisha McMath, was the justice of the peace and postmaster for the mining community. McMath and Colonel Tannehill were two of Bucksville's leading citizens and served as trustees for Bucksville Male and Female High School. The McMath Place, located to the west of town, was known for its large spring.

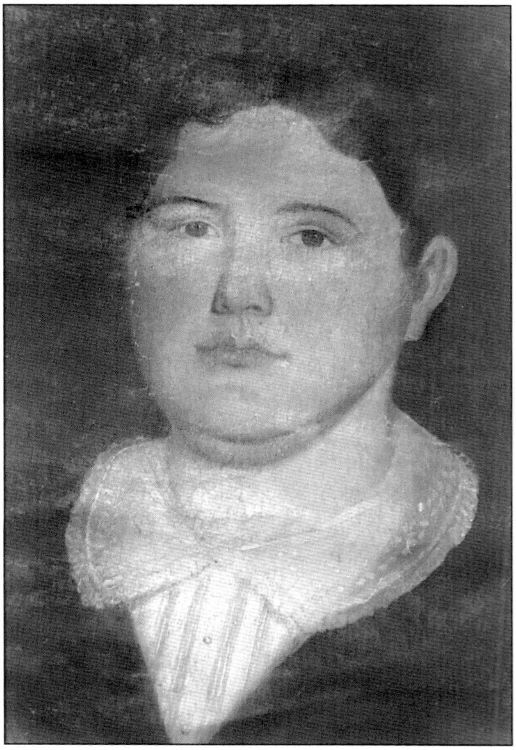

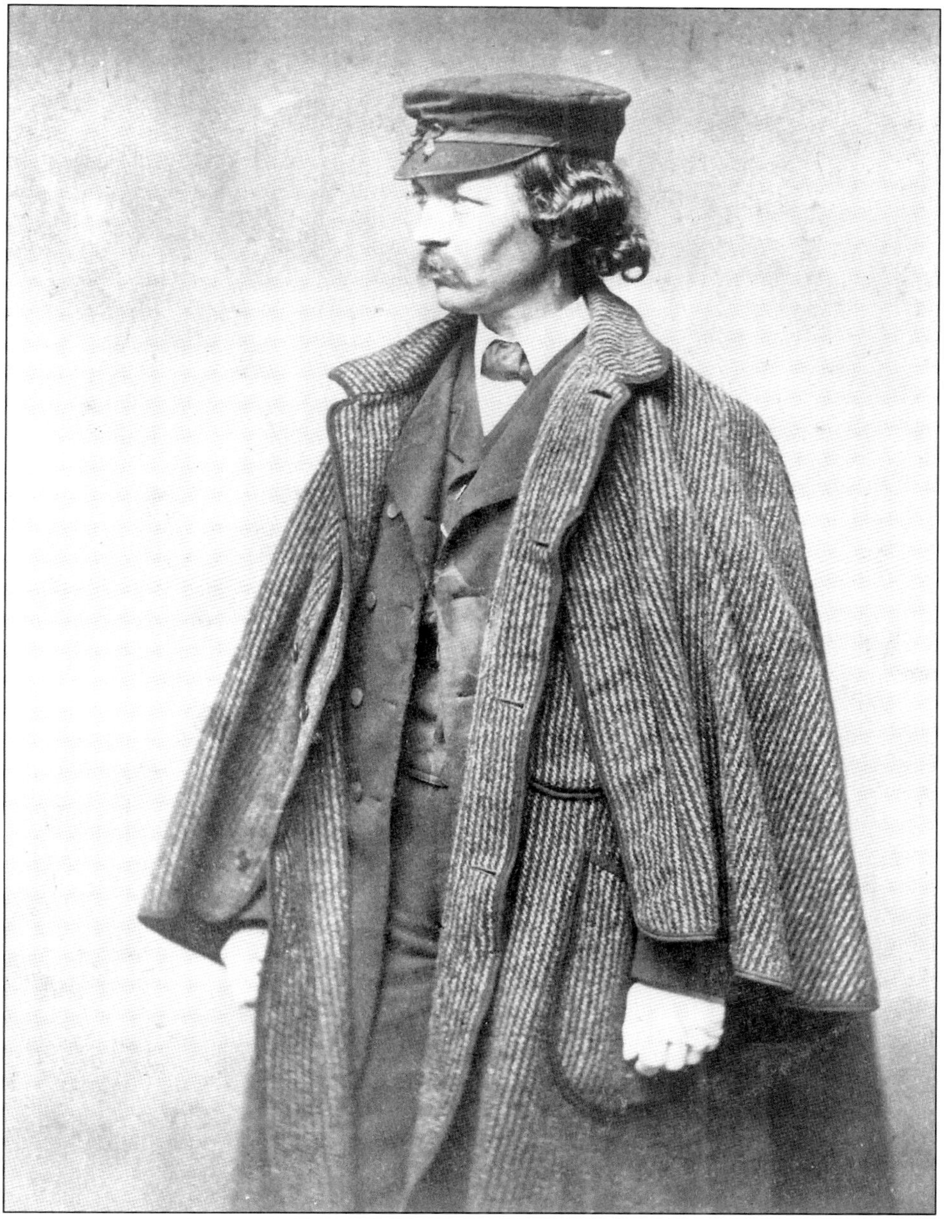

Noted landscape architect Frederick Law Olmsted, shown around 1860, was a visitor to the Tannehill forge in 1854. He described mining activities as being performed by entire families, some of whom were "powerfully poor and amazingly ignorant." Both slaves and Irishmen, he said, worked at the forge. Olmsted later designed Central Park in New York and the terraces for the US Capitol. Olmsted made recommendations on how to improve the Alabama State Capitol grounds in 1889 and drew a plan for a park system that included Red Mountain in Birmingham in 1924. Olmsted was the leading landscape architect of the post–Civil War generation and has long been acknowledged as the founder of American landscape architecture. Among his last projects were a park system for Boston and the design for the 1893 World's Fair in Chicago. (National Park Service, Frederick Law Olmsted National Historic Site.)

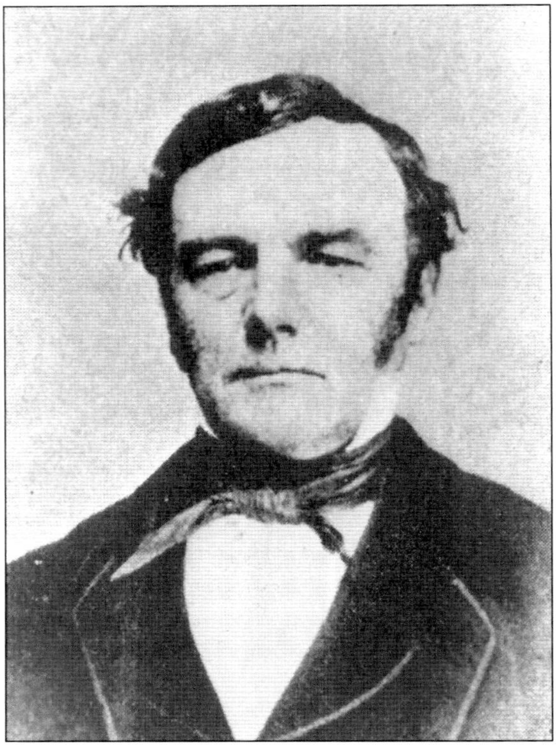

Daniel Hillman Jr., the son of the Tannehill forge builder, became an iron mogul in his own right as owner of the Empire, Center, and Fulton Furnaces in Kentucky. After the Civil War, he bought large land holdings near Birmingham that would later be passed on to his son and sold to the Tennessee Coal, Iron & Railroad Company. (Ethel Armes.)

Coal mine owner Benjamin F. Roden built the Hillman Hotel, shown around 1906, at the corner of Nineteenth Street and Fourth Avenue in downtown Birmingham. Named for T.T. Hillman, it opened in 1901 and became the center of Birmingham social life prior to World War I. The building was demolished in 1967. (Birmingham Public Library.)

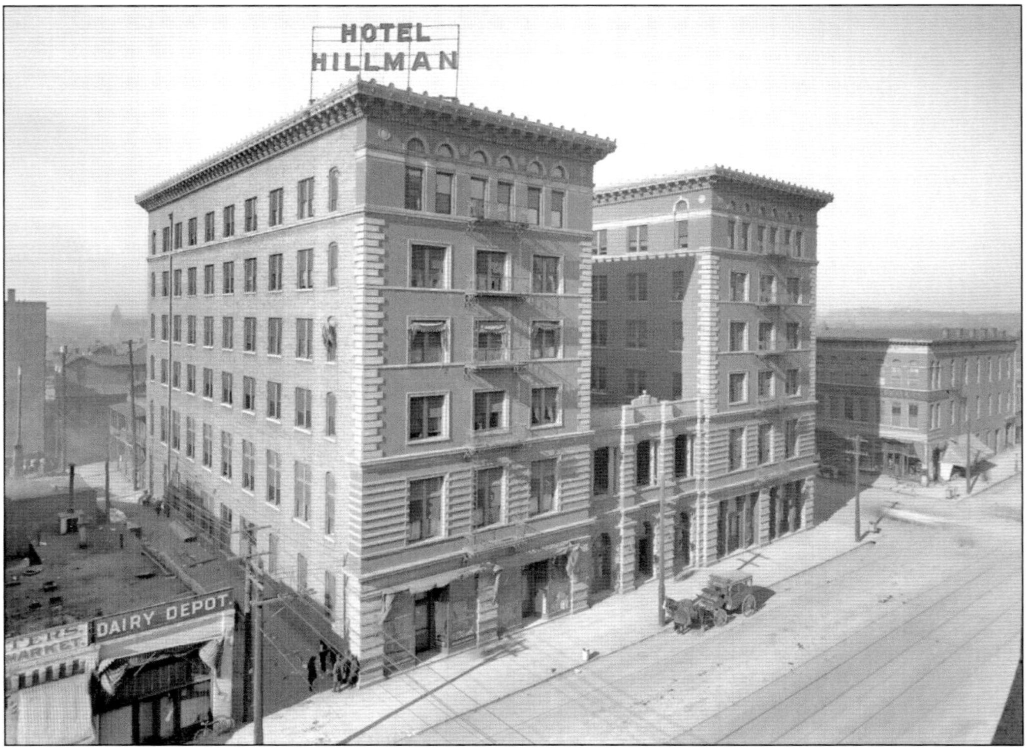

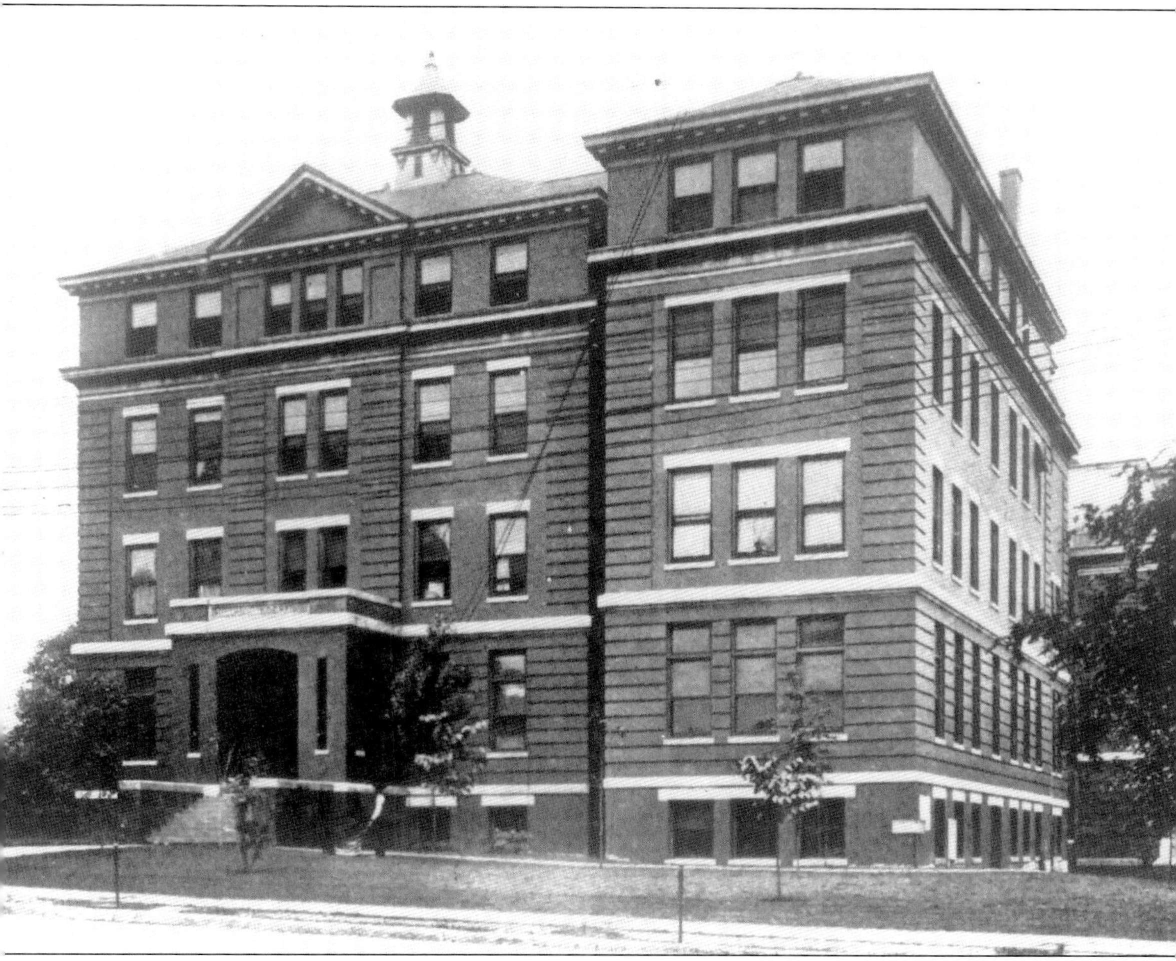

Hillman Hospital, built in 1903, was also named for T.T. Hillman. The clinic is still in use today as part of the University of Alabama Medical Center in Birmingham. Hillman supported the city's charity hospital during the 1890s and was honored for his service when the older facility was renamed Hillman Hospital in 1896. The community of Hillman, located west of Birmingham, was named for Hillman as well. (Birmingham Public Library.)

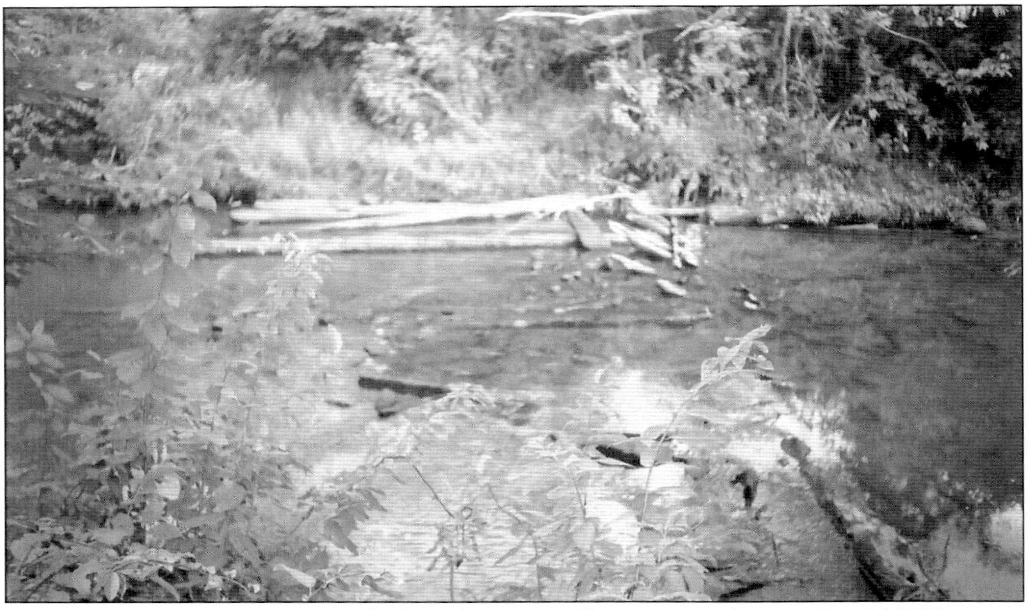

After over a century, remnants of a large log dam still remain in place at the old Williams & Owen Forge site a mile downstream from the Tannehill Ironworks on Roupes Creek. The dam powered a waterwheel that provided air blast and operated the forge machinery. An archaeological study in 2009 found the forge remains on the south side of the creek, most of them covered by dirt moved by preceding floods.

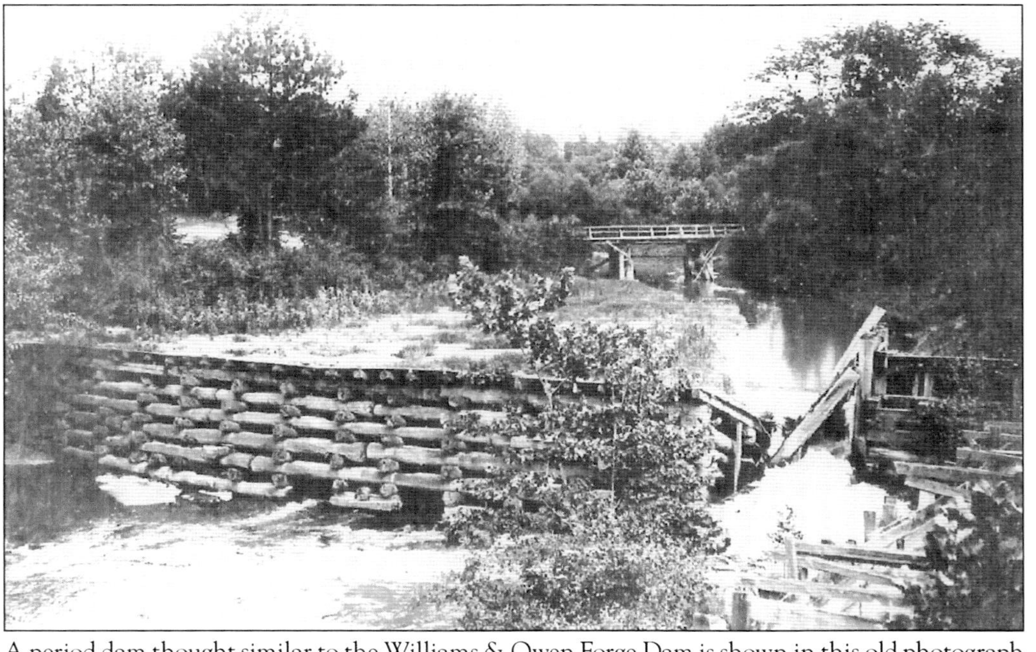

A period dam thought similar to the Williams & Owen Forge Dam is shown in this old photograph from another site in Bibb County around 1910. Dams of this type, built with logs held in place by rocks, were common along Alabama streams. The large hand-hewn log pieces are the most prominent part of the old forge still remaining at Tannehill. (Marty Everse, Birmingham, Alabama.)

Another view of the old Williams & Owen Dam shows large pieces of the structure still in place after 150 years. While much of the log dam has been washed downstream by floods, significant pieces show the detailed workmanship that went into notching it together. Among the artifacts found at the site is a piece of the winged gudgeon that was attached to the waterwheel.

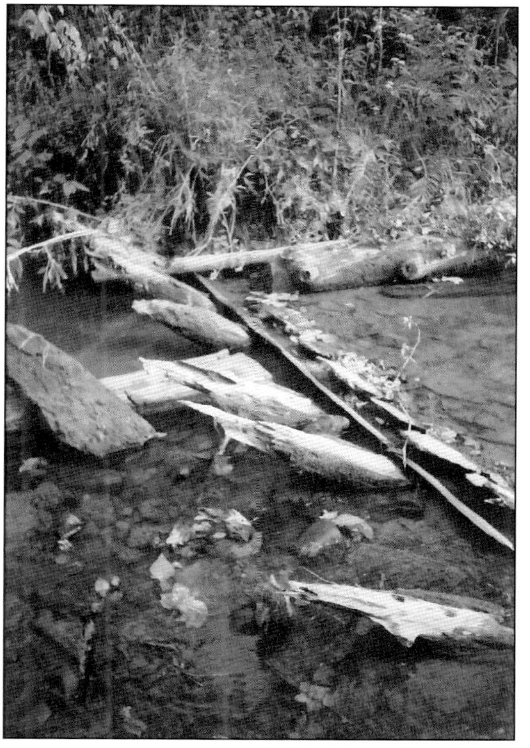

The home of Thomas Hennington Owen on Eastern Valley Road in Bessemer was built in the 1830s by the co-owner of the Williams & Owen Forge at Tannehill. The bloomery operated from 1861 through the Civil War and was destroyed by a spring flood in 1866. Owen was a farmer and merchant who also served as a Jefferson County commissioner in the late 1870s. (Bessemer Hall of History.)

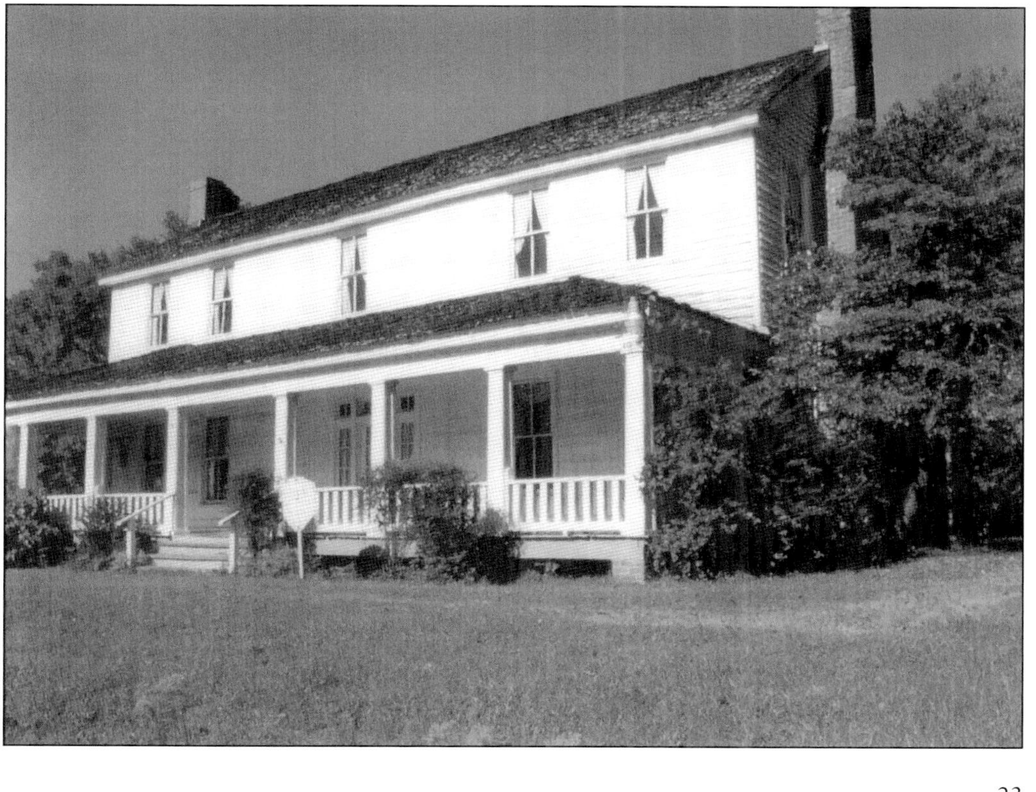

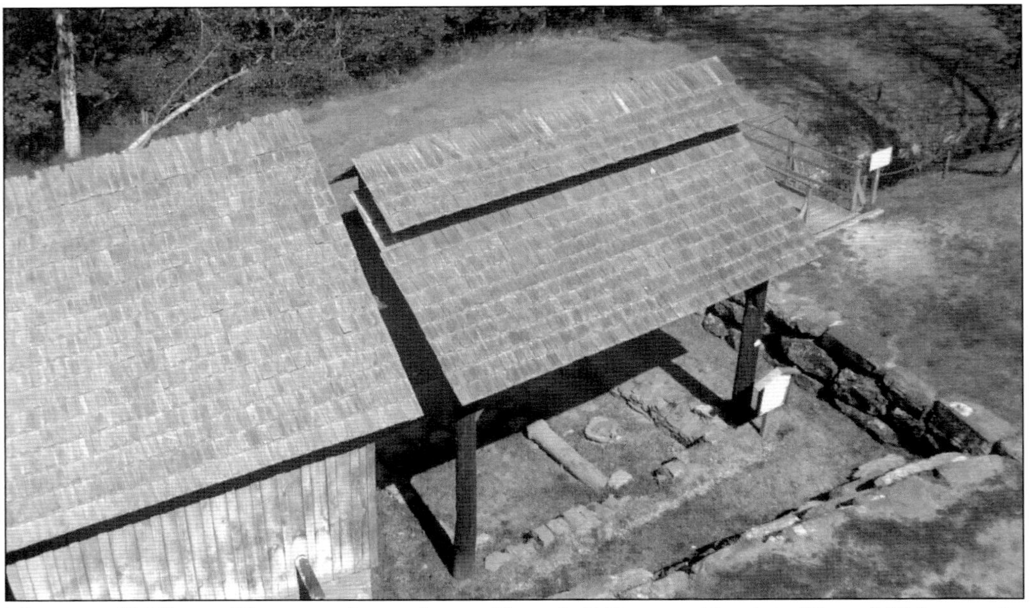

The site of Hillman Bloomery, located near Tannehill Furnace No. 1 in Tannehill Ironworks Historical State Park, shows remnants of an old forge. Referred to by historian Ethel Armes as the "Old Timey Forge," this site may have been converted into a refinery forge during the Civil War. (John Morse/Bhamwiki.)

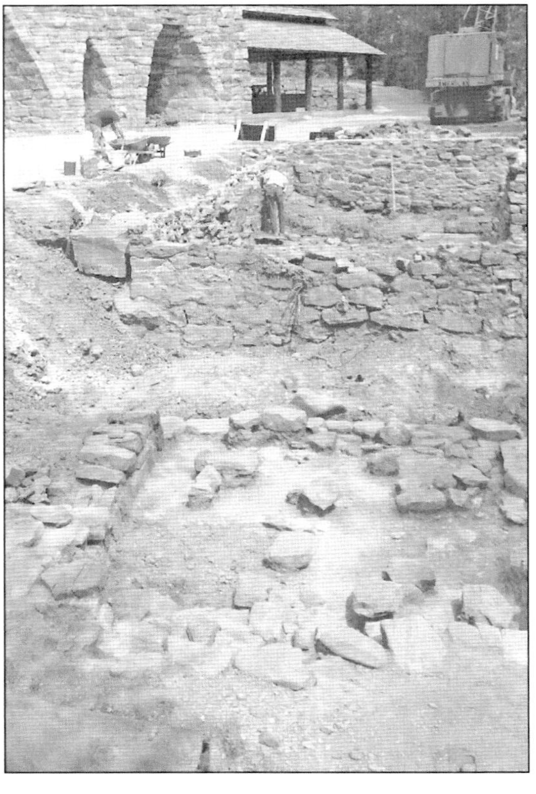

This bird's-eye view of the University of Alabama archaeological investigation at the Tannehill Furnace site in 1990 reveals what is thought to have been the blower house and forge foundations for the Hillman Bloomery. A wooden duct pipe, however, runs toward the ruins of another suspected forge site nearer Furnace No. 1. What appears to be a ditch just out of view to the right is the forge water race, which was known to have been in place by 1842. (*Birmingham News.*)

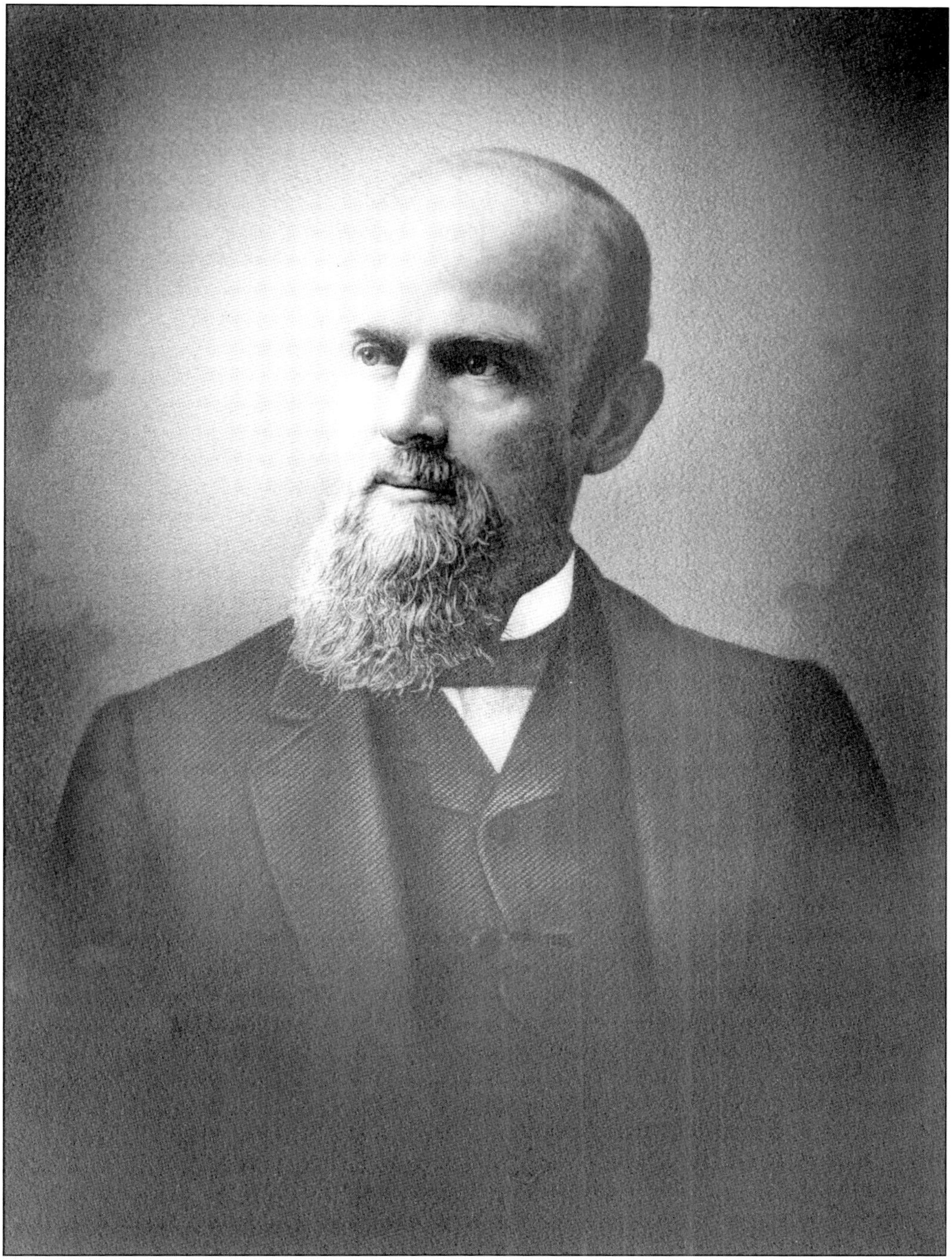

Thomas T. Hillman, Daniel's son and the grandson of the forge builder, would build upon his inheritance to become one of Alabama's leading iron and steel executives. He became vice president of the Tennessee Company, which was sold in 1907 to US Steel. In 1884, he also helped build the first blast furnace in the city of Birmingham, the Alice Furnace, of which he became president. (Birmingham Public Library.)

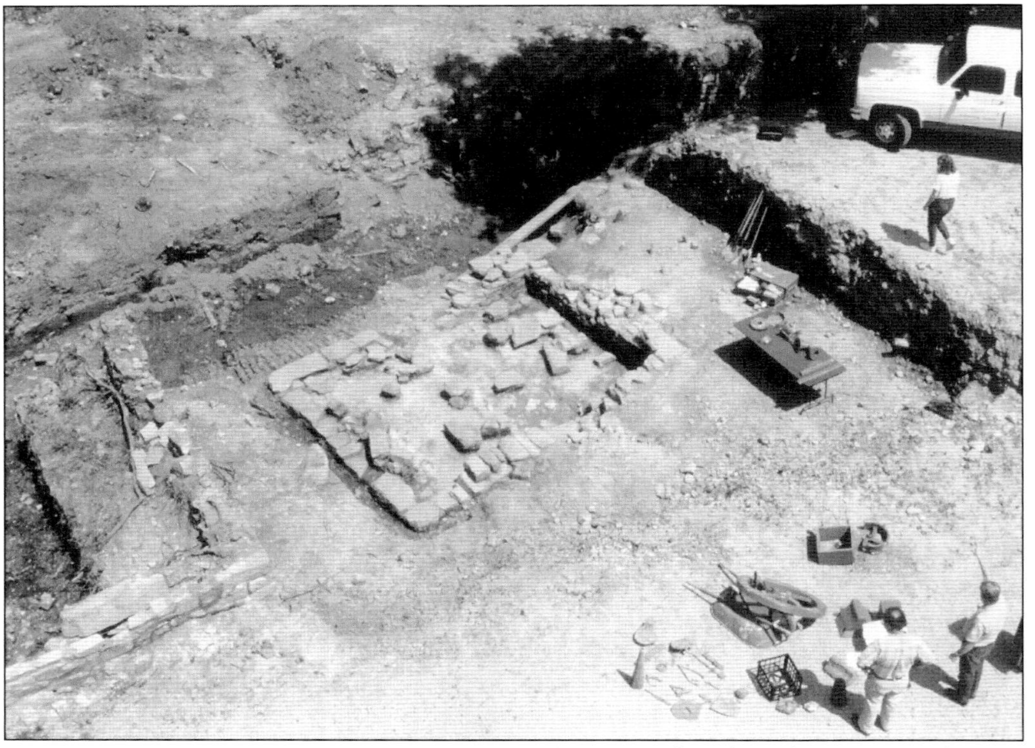

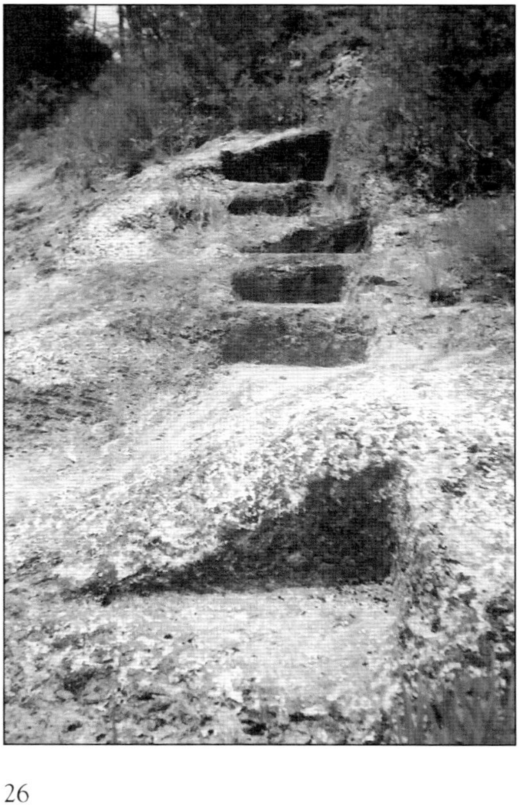

A major archaeological excavation of the Tannehill Forge and Furnace area was conducted by the University of Alabama in 1990 under the guidance of archaeologist Carey Oakley. The work led to the restoration of both the forge site (1830) and the blower house for Tannehill Furnace No. 1 (1859). A forebay or canal (1840) on which the waterwheel was located is also preserved.

William P. Browne built a dam in 1847 on the Little Cahaba River about 150 yards upstream from the site of the Brighthope Bloomery to provide it with air blast. The forge, which went into blast in the next year, was expanded with an even larger plant in 1864. The stone steps on which the log dam sat are still in place.

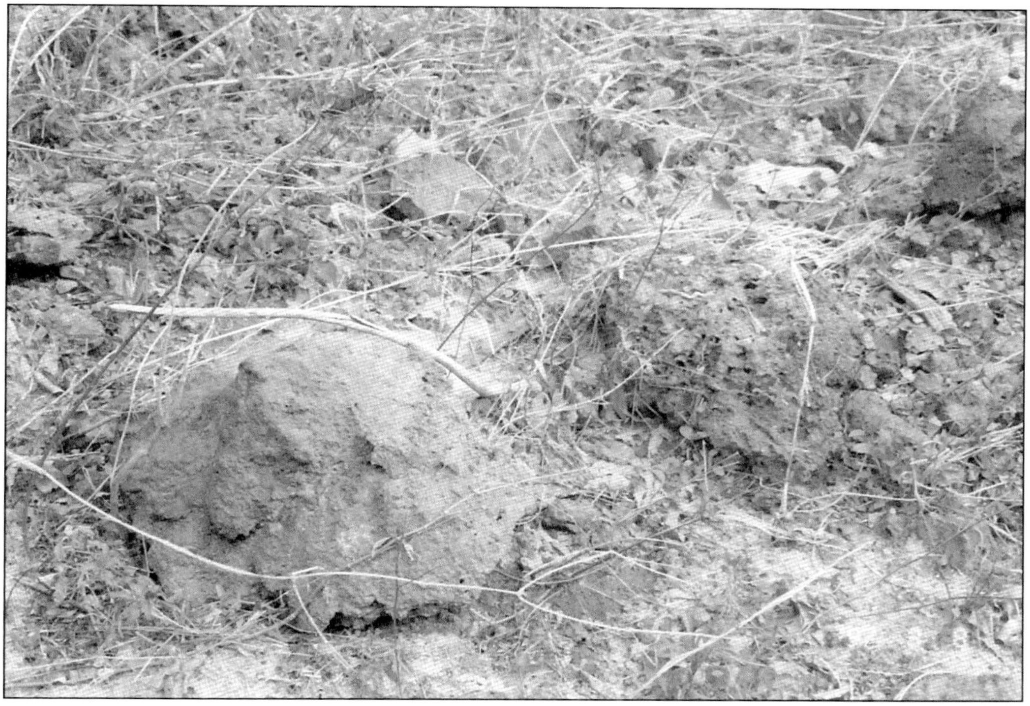

"Skulls," a rounded waste product from a forge run, still mark old forge sites around the state, including this photograph taken at Maria Forge near Waldo in Talladega County. E.R. and Samuel Riddle built Maria, one of the largest of the Alabama bloomeries, in 1843. The inefficient ore reduction process used by these early forges left a high iron content in the slag.

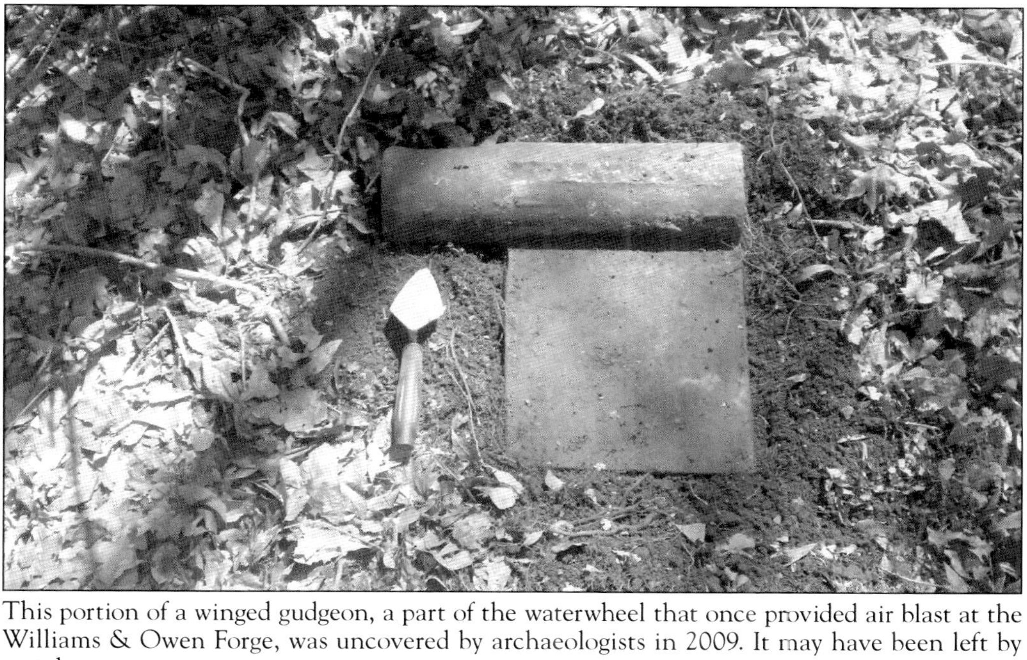

This portion of a winged gudgeon, a part of the waterwheel that once provided air blast at the Williams & Owen Forge, was uncovered by archaeologists in 2009. It may have been left by metal scavengers.

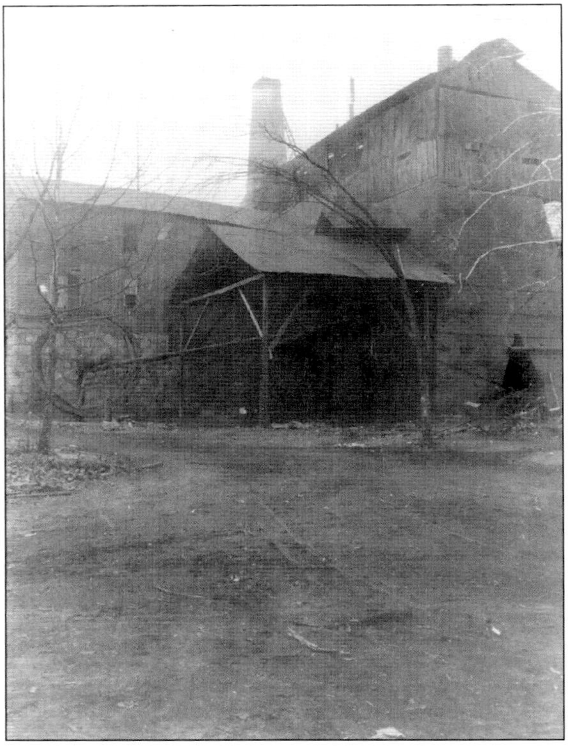

Here is a c. 1908 photograph of Center Furnace, built in 1844 by Daniel Hillman Jr. in Trigg County, Kentucky. The furnace was the largest in the area and continued in operation until it closed in 1912. Many of the laborers working at this furnace were Chinese immigrants. The furnace was made of brick rather than stone.

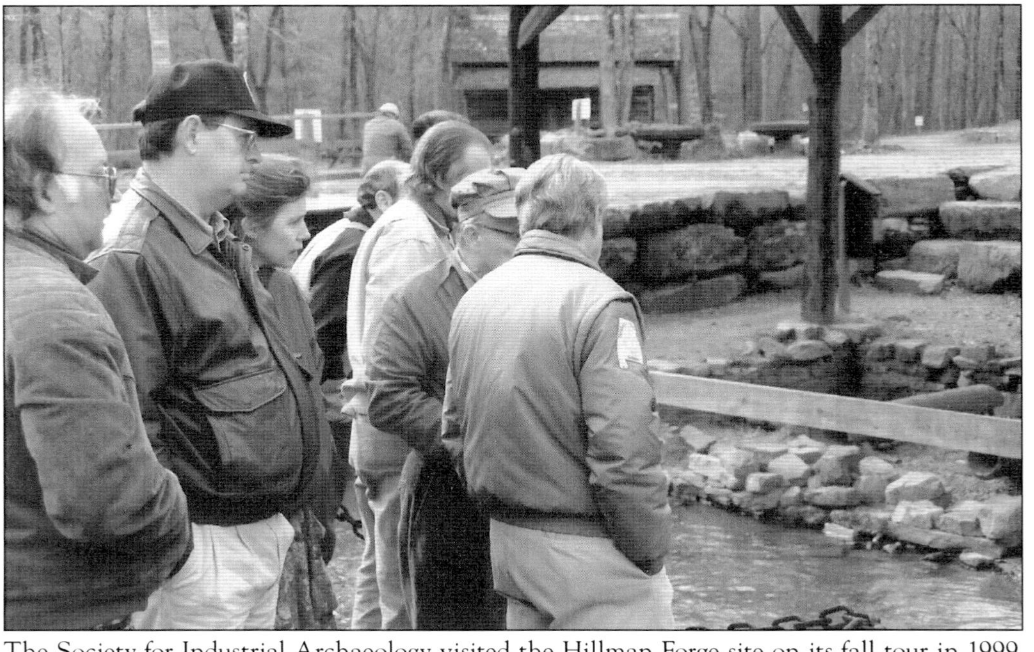

The Society for Industrial Archaeology visited the Hillman Forge site on its fall tour in 1999, bringing many national authorities on early forges to Birmingham. The forge ruins, along with the nearby Tannehill Furnaces, are listed on the National Register of Historic Places and the American Society for Metals listing of international landmarks.

Two

Stroup's Blast Furnaces

Moses Stroup was thought of by many of his contemporaries to be the South's leading iron maker. At various times in his career, he managed the largest ironworks in South Carolina, built the first rolling mill in Georgia, and made some of the first rail in the South.

In Alabama, his wartime experiments with Red Mountain iron ore paved the way for construction of the first blast furnaces in Jefferson County, which was where Birmingham would emerge a few years later.

A partner in the Etowah Ironworks near Cartersville, Georgia, in the 1840s, Stroup left the successful enterprise to move to Alabama, where he believed iron could be made more cheaply than anyplace else in the country. Interestingly, the town of Cartersville, Georgia, was originally named Birmingham.

Stroup began construction of the Round Mountain Furnace, near Centre, in Cherokee County in 1848. When faced with financial difficulties, he headed to Tuscaloosa County, where he was invited by Col. Ninian Tannehill, the owner of a plantation forge, to expand the works by adding a blast furnace.

Tannehill sold the forge in 1857 to John Alexander, an old Stroup acquaintance from South Carolina who ran a large farm and gristmill near Montevallo. With Alexander's financing, the first blast furnace at Tannehill went into blast some time between 1858 and 1859 and produced six to eight tons a day. With the outbreak of the Civil War, work began on two more furnaces in 1862, funded by the Confederate government for military needs. This battery was completed by William L. Sanders after Stroup left on Confederate contract to build new furnaces at Oxmoor in Jefferson County.

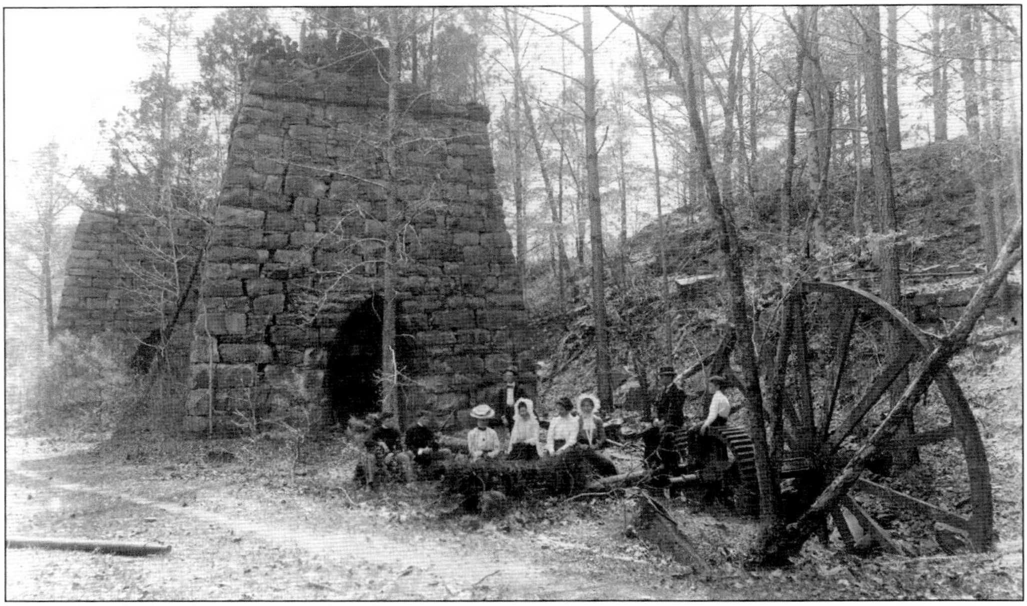

This photograph, taken by state geologist Eugene A. Smith for the Alabama Geological Society in 1905, reveals an outing at the Tannehill Ironworks site amid old gearing, blast pipe, and the steam engine's 15-foot flywheel. A few brave visitors in this view have climbed atop Furnace No. 3, where the brick draft stack was then still partially in place. (Alabama Department of Archives & History.)

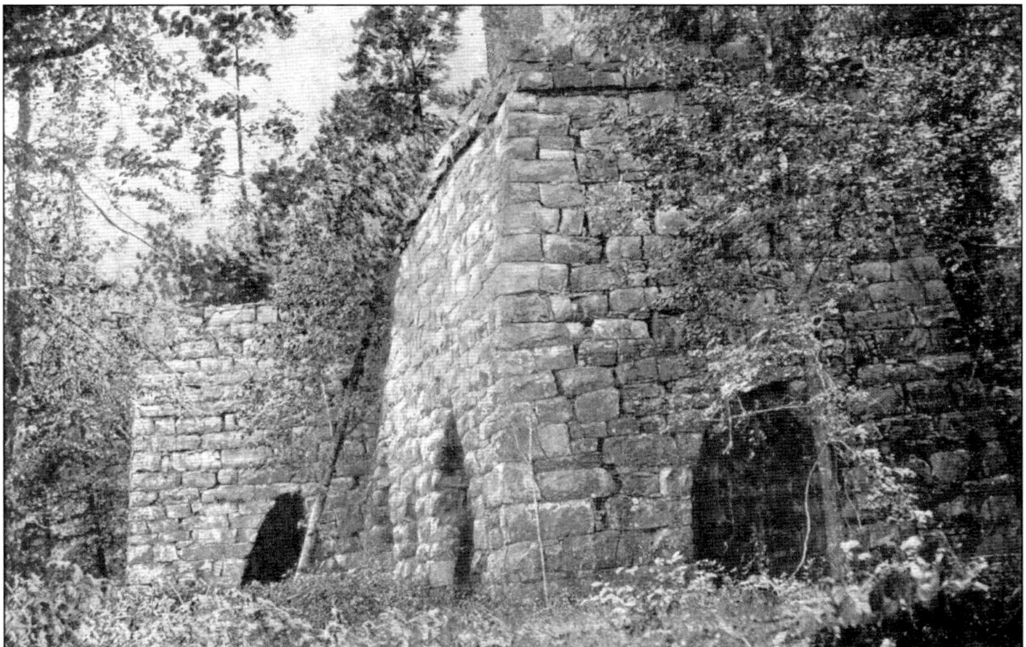

The old Tannehill Furnaces, as they appeared about 1910, show the ravages of time and weather. This front portion of the double furnaces, erected in 1862 with Confederate funding, is beginning to show the fatal crack in its stone face that resulted in a major collapse in 1930. (Ethel Armes.)

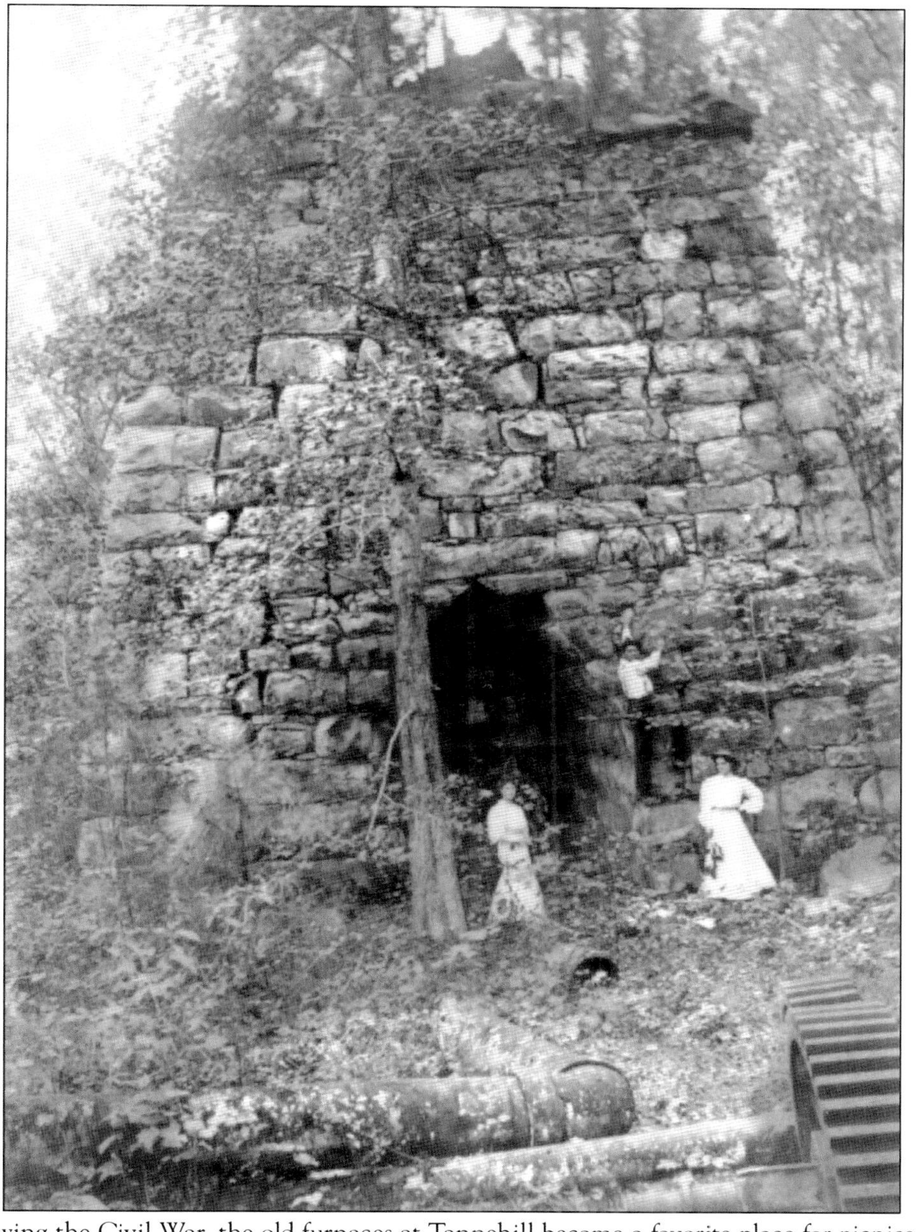

Following the Civil War, the old furnaces at Tannehill became a favorite place for picnics and outings. In this photograph, air blast pipe and gears mark the area where the mill's steam engine was located in the late 1890s. The plant's steam engine had a 15-foot flywheel. It was so large that a portion of it was located in a rock-lined trench in the ground. The engine provided air blast to expedite ore reduction in Furnaces Nos. 2 and 3. Hot blast stoves were initially installed in the operation but may have later been discontinued. For years, old pieces of broken machinery and air pipe littered the furnace yard, but they gradually disappeared as metal scrappers made raids on the site, despite efforts of the Republic Iron & Steel Company to stop the practice. Scrap drives in both world wars caused many rare pieces of industrial machinery here to disappear. (Harper Collection, Samford University Library.)

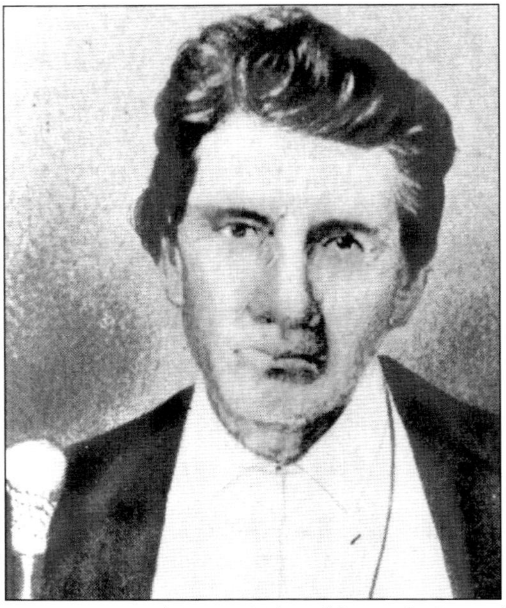
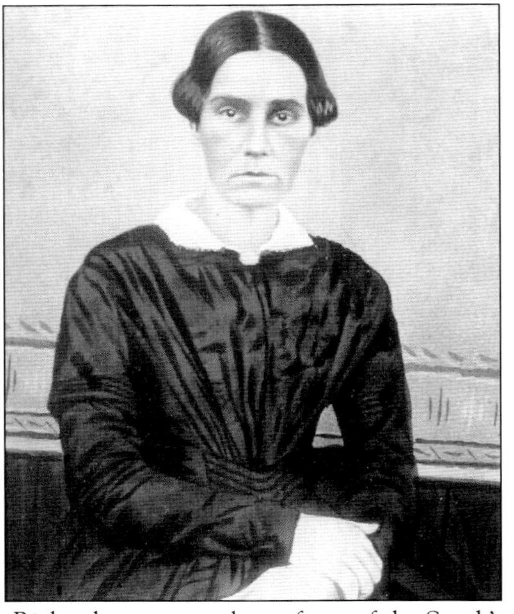

Moses Stroup (1794–1878) and his wife, Permelia Richards, were members of one of the South's leading iron-making families. Stroup moved from North and South Carolina into Georgia and Alabama. Permelia, Stroup's second wife, was 19 years his junior. Born in South Carolina in 1813, she is thought to have died between 1850 and 1855, presumably in Bartow County, Georgia, near the Etowah Furnace, before he came to Alabama. (Robert Ausman, Hueytown, Alabama.)

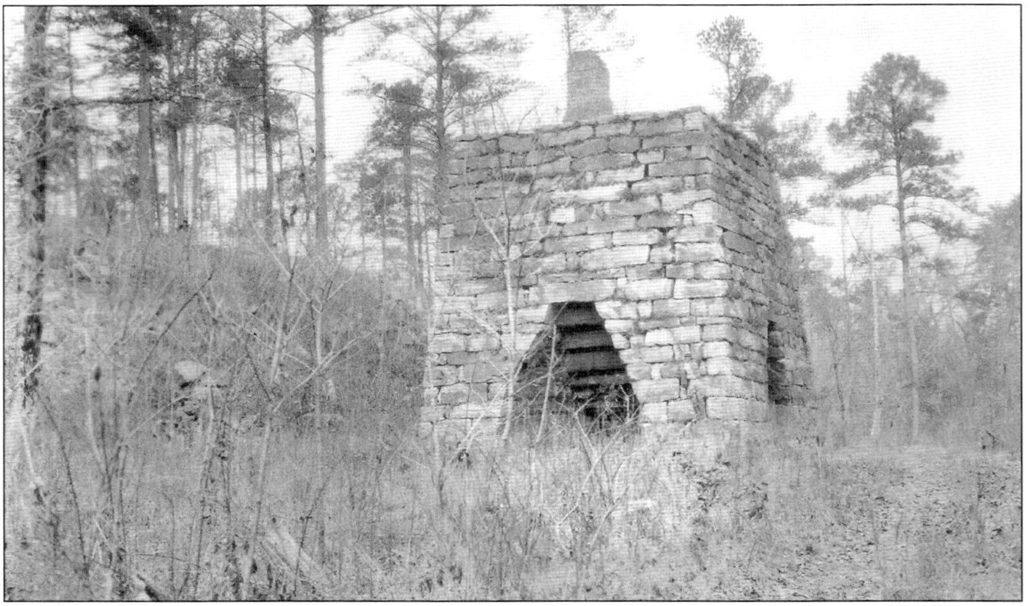

Stroup's first furnace at Tannehill was erected in 1859, which was some four years after he arrived in Roupes Valley to upgrade the old Hillman Forge on the plantation of Ninian Tannehill. The furnace, which fell in 1862 but was quickly rebuilt, could produce six to eight tons of pig iron each day. This photograph was taken as it appeared in 1929. (E.F. Burchard photograph, US Geological Survey Photographic Library.)

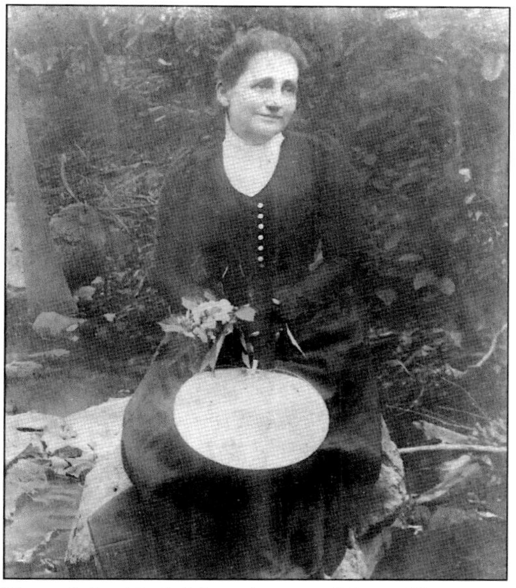
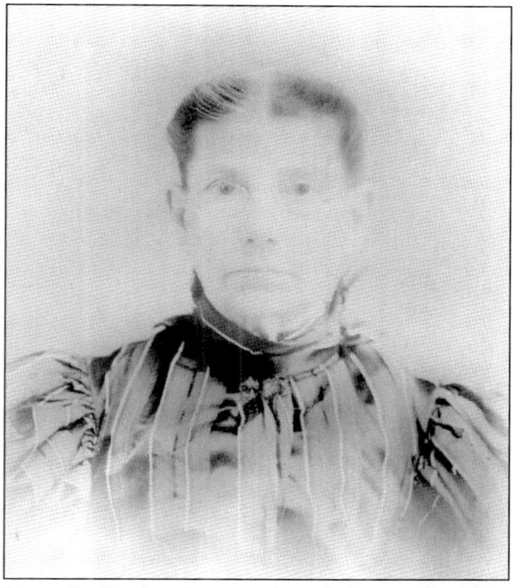

Two of the 10 children born to Moses Stroup and Permelia Richards were, from left to right, Ellie and Susan. Ellie, also known as Margaret Ellen Manring, was born in 1849 and died in 1931 in Tuscaloosa. Susan Wilmoth Stroup was born in 1834 and died in 1904 in Ensley, Jefferson County. She married Pleasant A. Hickman in 1862. Susan was born in South Carolina, and Ellie was born in Georgia. (Robert Ausman.)

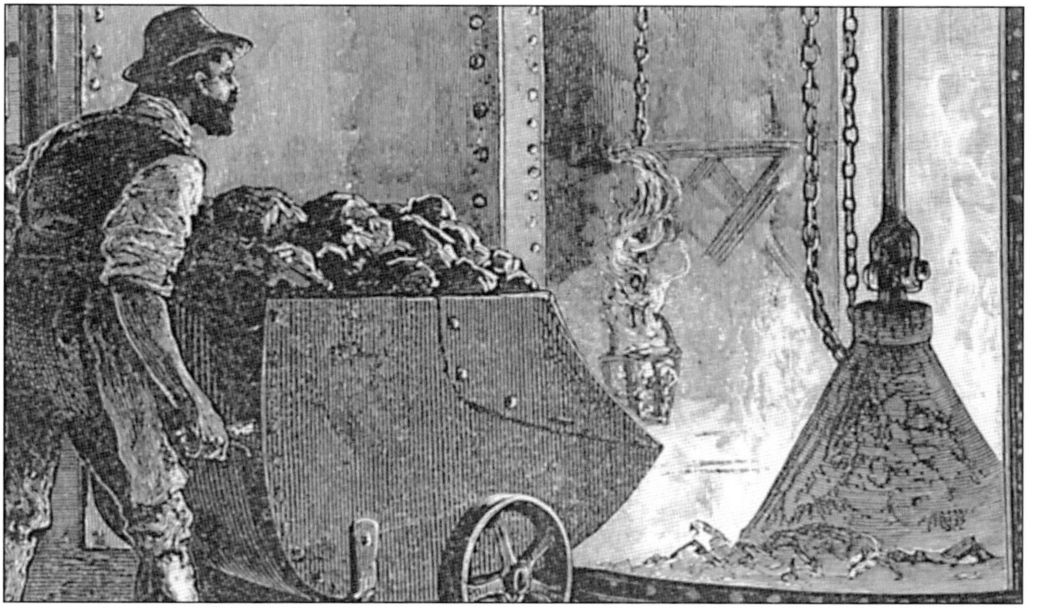

This is an illustration of a worker charging a bell and hopper around 1880 as shown in C.L. Mateaux's *Wonderland of Work*. The bell and hopper was a device used to evenly disperse loads of iron ore and limestone in a top-fed blast furnace. While the first bell and hopper in Alabama was installed at the Shelby Iron Company, a bell and hopper, also called a filler and thimble, was used at Tannehill in the double furnaces and was listed in a postwar sheriff's sale of disposable equipment in 1865. The first bell and hopper was invented by George Parry in England in 1850.

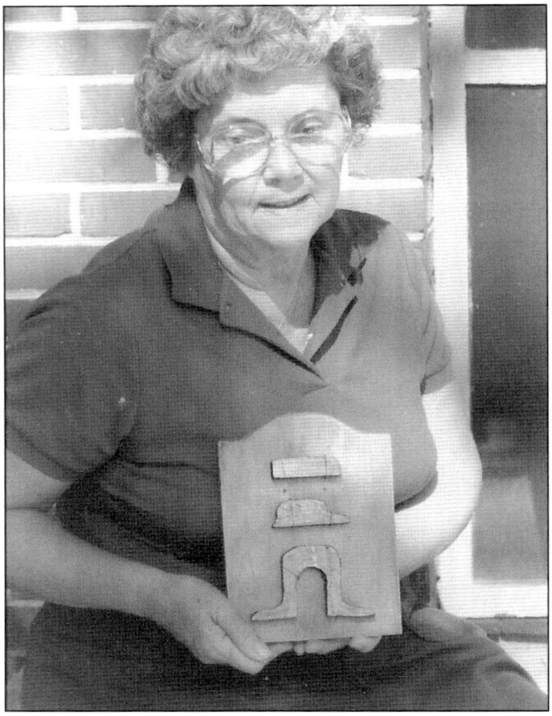

Jane Maxwell, the great-great-granddaughter of Moses Stroup, holds pieces of iron rails cast by the old ironmaster while at Cartersville, Georgia. Stroup made some of the first rails in the South at his Cass County rolling mill. Known as iron strap, it was first used on the State Line, later known as the Western & Atlantic. The rolling mill was added to the Old Etowah Furnace in 1846 in partnership with Mark A. Cooper.

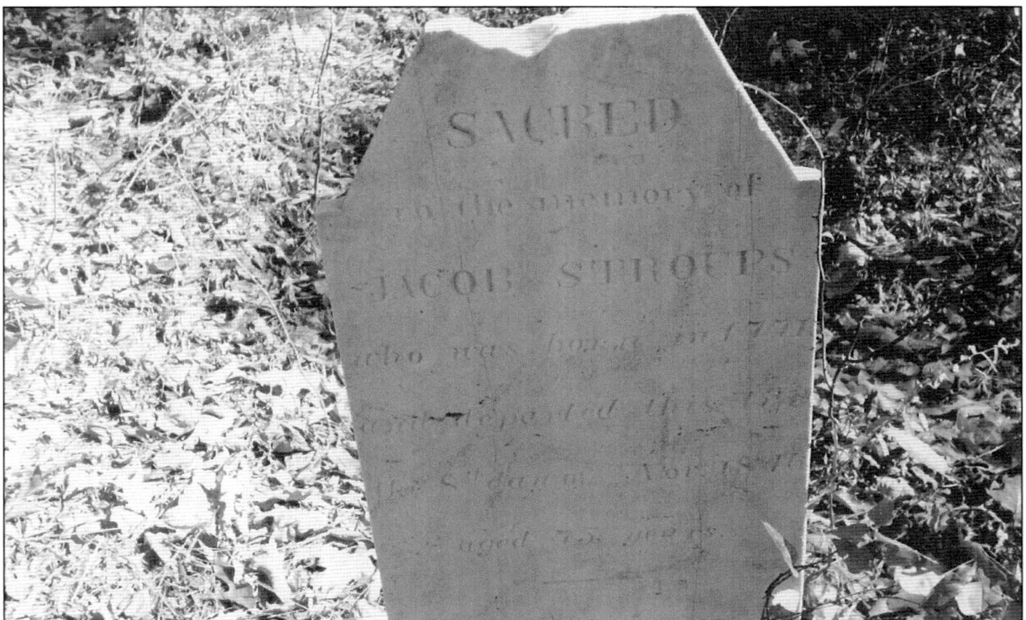

Jacob Stroup, Moses's father, built blast furnaces in South Carolina, Georgia, and Alabama, including the Cane Creek Furnace in Calhoun County in 1840. Jacob returned to Georgia to erect another furnace on the east bank of Allatoona Creek, near Cassville, and died there at age 75 in 1846. He is buried at the Goodson and Rogers Cemetery, which is located across the road from his last home. (John R. Jackson, Dallas, Georgia.)

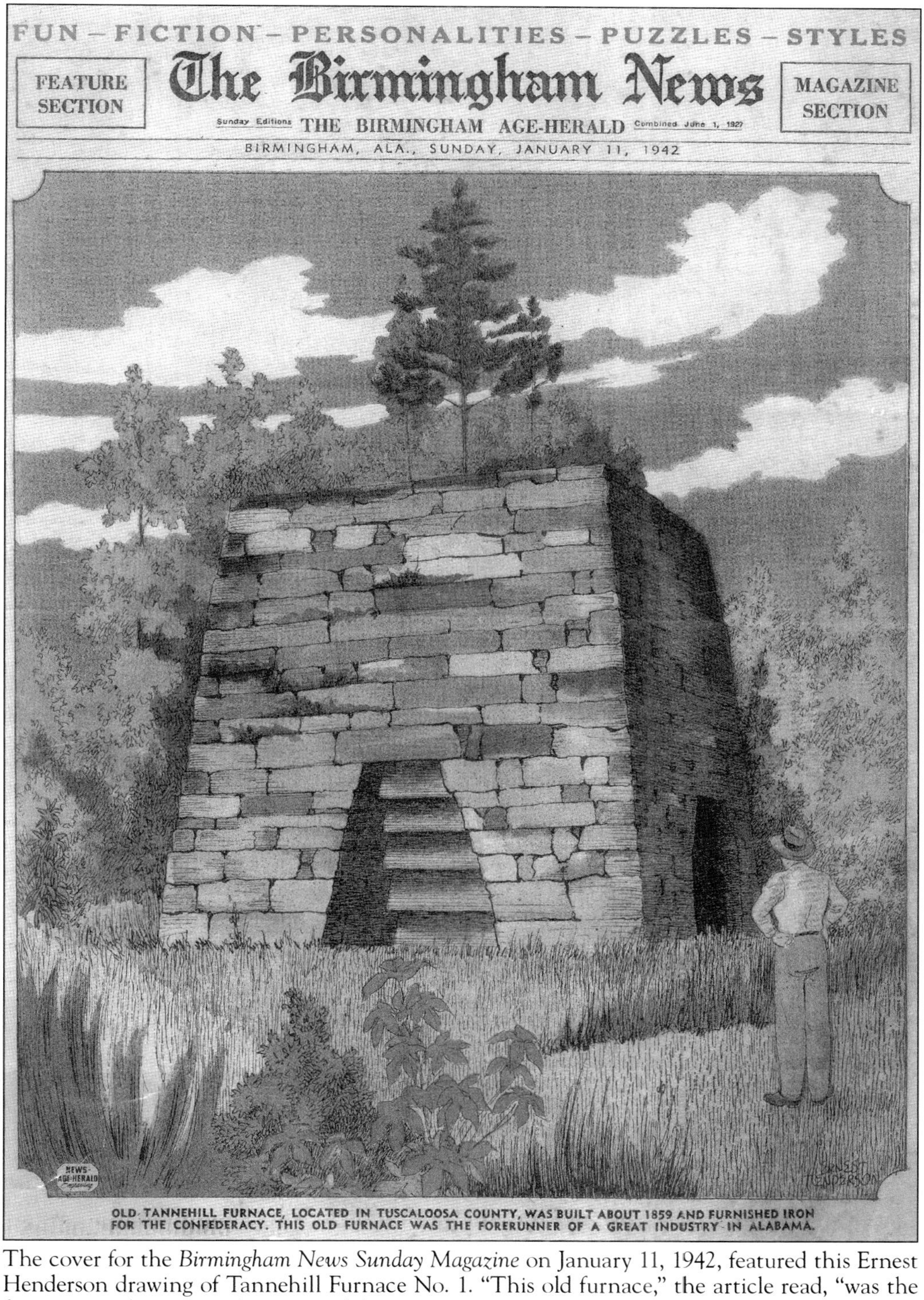
FUN – FICTION – PERSONALITIES – PUZZLES – STYLES

FEATURE SECTION

The Birmingham News

MAGAZINE SECTION

Sunday Editions THE BIRMINGHAM AGE-HERALD Combined June 1, 1927

BIRMINGHAM, ALA., SUNDAY, JANUARY 11, 1942

OLD TANNEHILL FURNACE, LOCATED IN TUSCALOOSA COUNTY, WAS BUILT ABOUT 1859 AND FURNISHED IRON FOR THE CONFEDERACY. THIS OLD FURNACE WAS THE FORERUNNER OF A GREAT INDUSTRY IN ALABAMA.

The cover for the *Birmingham News Sunday Magazine* on January 11, 1942, featured this Ernest Henderson drawing of Tannehill Furnace No. 1. "This old furnace," the article read, "was the forerunner of a great industry in Alabama." It was the most complete of the three furnaces in the Tannehill battery during World War II. (*Birmingham News*.)

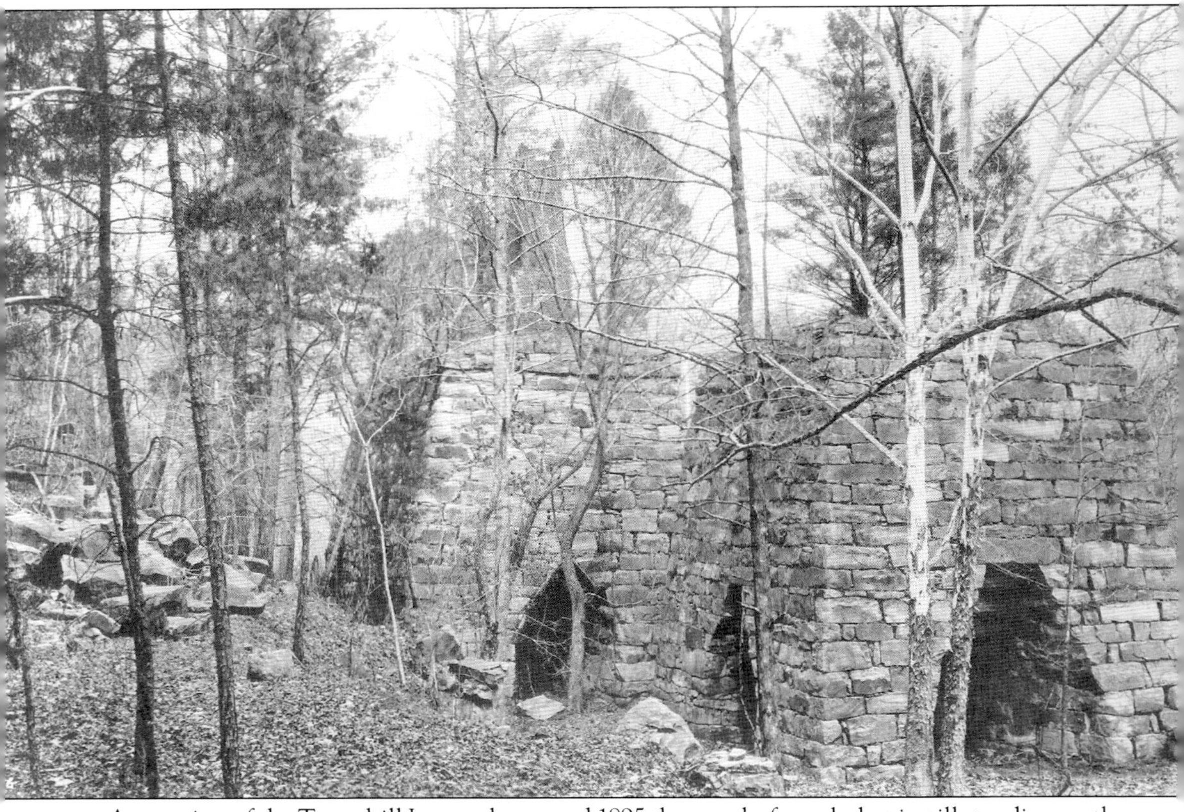

A rear view of the Tannehill Ironworks around 1895 shows a draft stack that is still standing on the double furnaces. Stroup's first furnace, built in 1859, is missing its stack. It was probably taken by brick scavengers years after the Civil War left them in shambles. An old furnace fire ring is seen in the lower right-hand corner. The rock pile to the left was probably associated with the furnace charging bridge, which was burned during the war. A backwoods curiosity, the industrial ruins became a favorite haunt for picnics and day-trips. The burned-out works at the time this picture was taken were under the ownership of the Republic Iron & Steel Company, which operated the Thomas Works in Birmingham. It continued to mine nearby deposits of brown ore while running relic hunters off the furnace property. (Birmingham Public Library.)

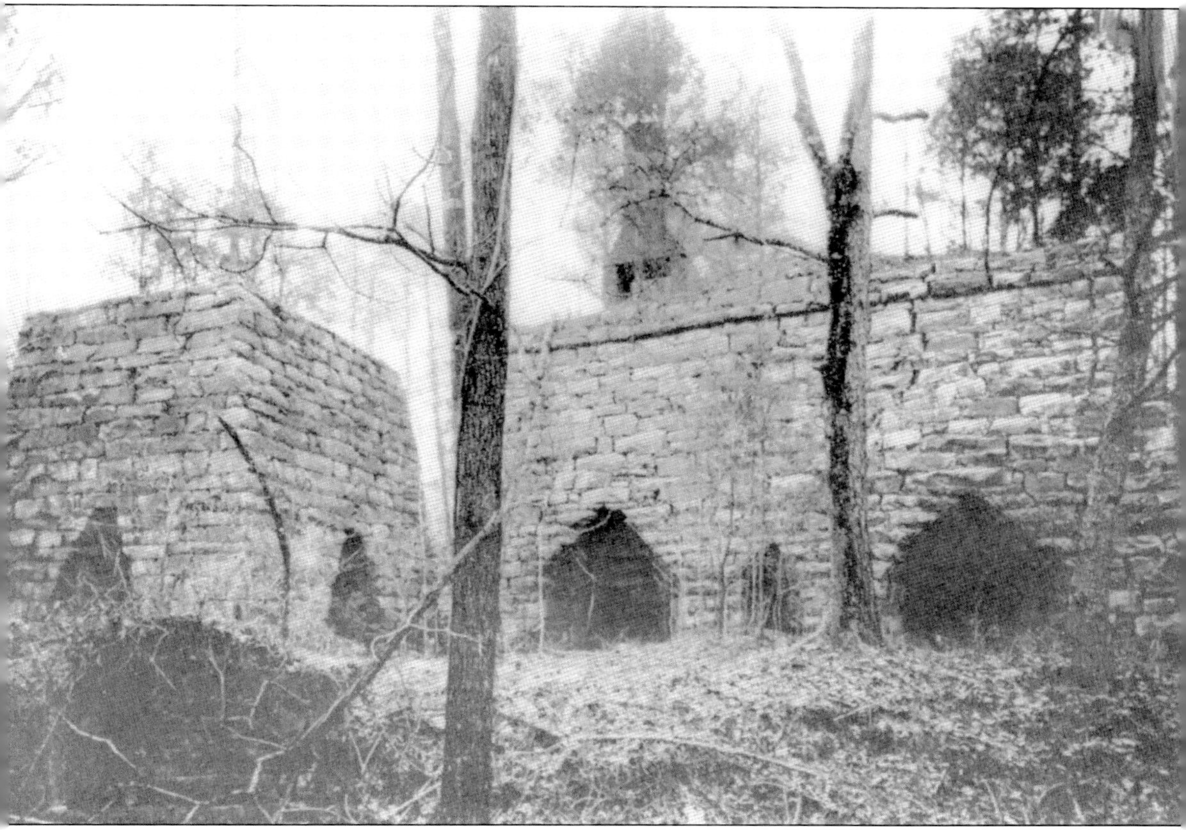

The Civil War–era blast furnaces at Tannehill were known as the Roupes Valley Ironworks and could produce 22 tons of iron a day for Confederate military needs. This photograph, dated 1922, shows all three furnaces still standing. The brick draft stack can be seen atop Furnace No. 2. The squared hole at the bottom of the chimney was the fitting for a hot blast stove. At the time, the idled ironworks was the property of the Republic Iron & Steel Corporation. This photograph was used inside the American Chemical Society's spring meeting program in Birmingham on April 3 through 7, 1922. The national conclave drew delegates from across the nation. (American Chemical Society.)

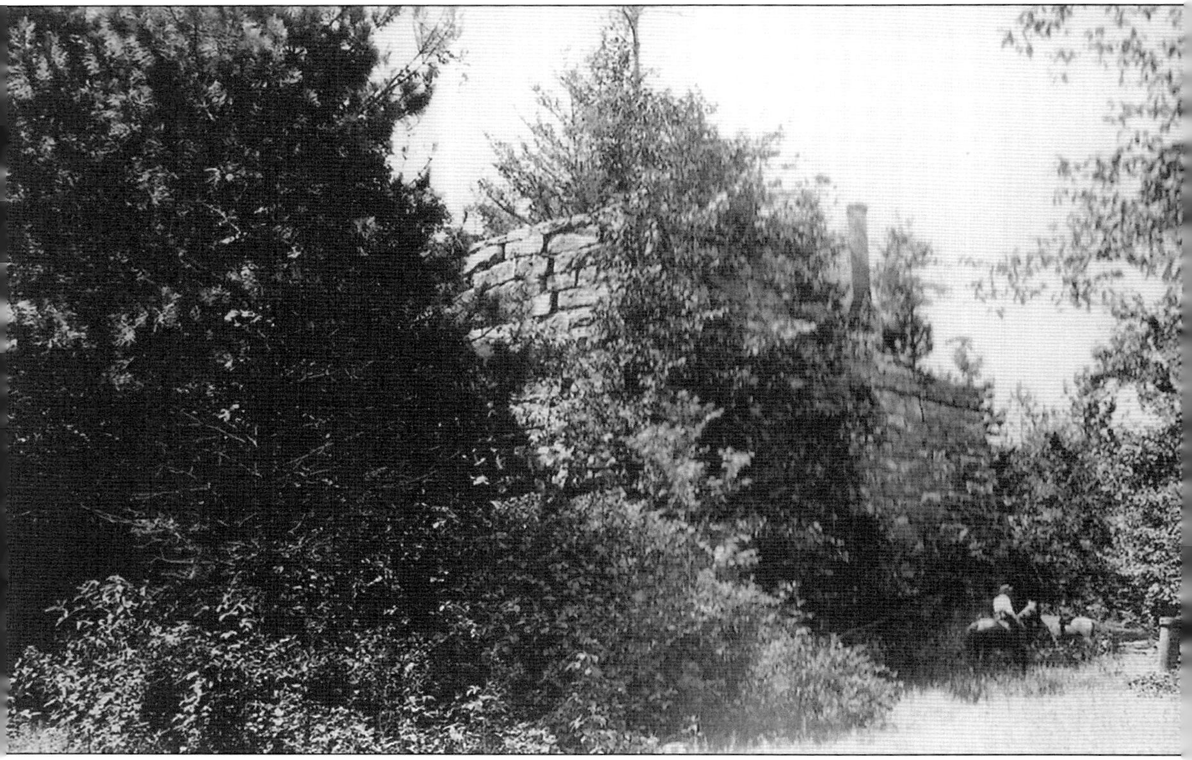

In this haunting picture from the past, probably from the 1890s, two visitors on horseback pause at the furnace site to take in its history. The photographer has left his mount to take the picture. Note the tall brick stack on the double furnaces with an apparent hood on top. This is the most complete picture of the stack known to exist.

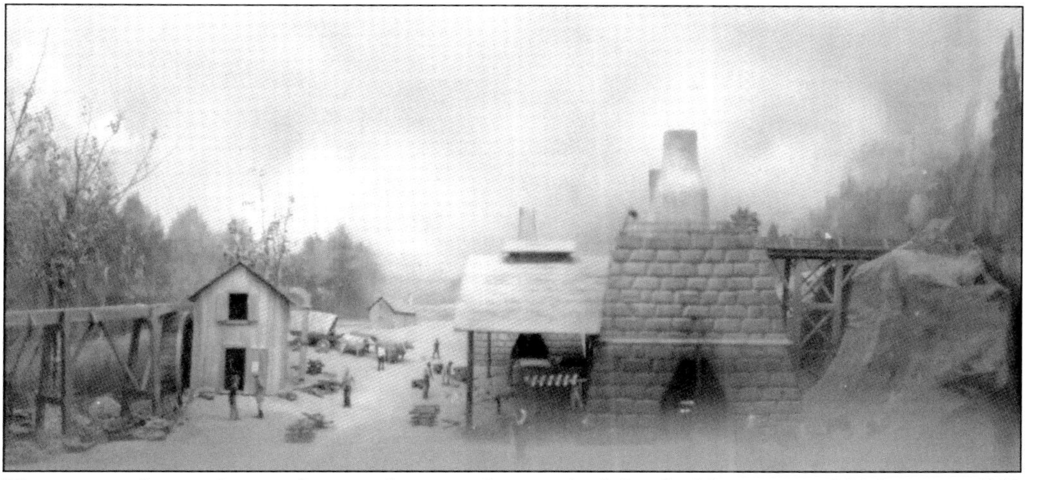

This postcard view shows a furnace diorama that was built by the University of Alabama in 1953 in preparation for an archaeological study three years later. The academic version of the furnace yard fails to take into consideration the need to have the raw materials for making iron—ore, limestone, and charcoal—protected from the weather. The charging bridge also would have needed a roof. (Scenic South Card Company.)

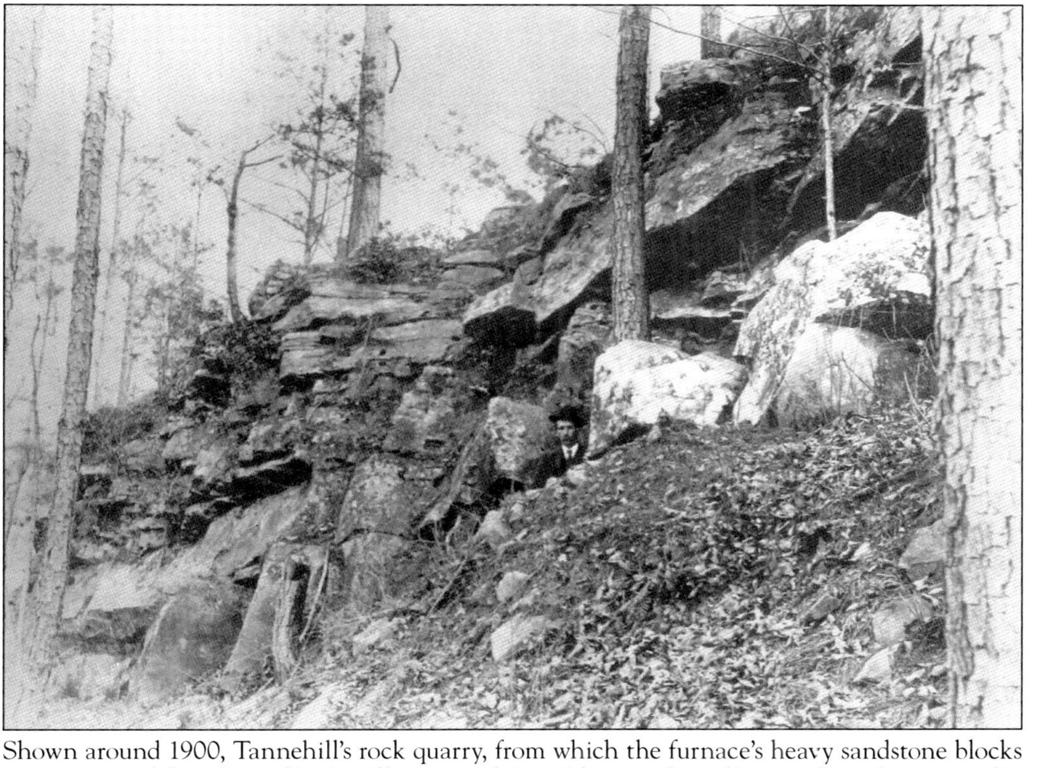

Shown around 1900, Tannehill's rock quarry, from which the furnace's heavy sandstone blocks were extracted decades earlier, is still very much in evidence a short distance from the ironworks. Some of the stones weigh in excess of 800 pounds and are thought to have been transported to the furnace site by use of skids. Slave stone cutters used star drills to cut the huge building blocks. (Birmingham Public Library.)

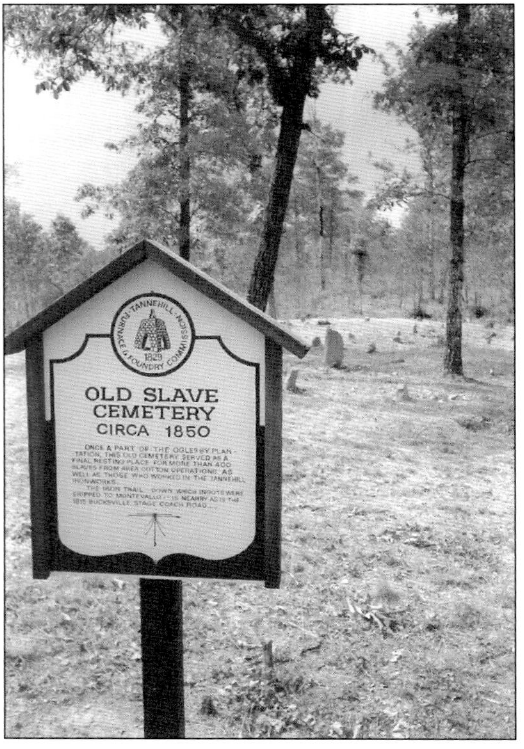

The slave cemetery at Tannehill State Park is one of its most haunting features. Containing the graves of more than 60 persons, some thought to be furnace workers, the site was discovered by Boy Scouts when clearing park land in the early 1970s. The cemetery is located along the Iron Haul Trail, the wagon route from the furnaces to the railhead at Montevallo, 18 miles away.

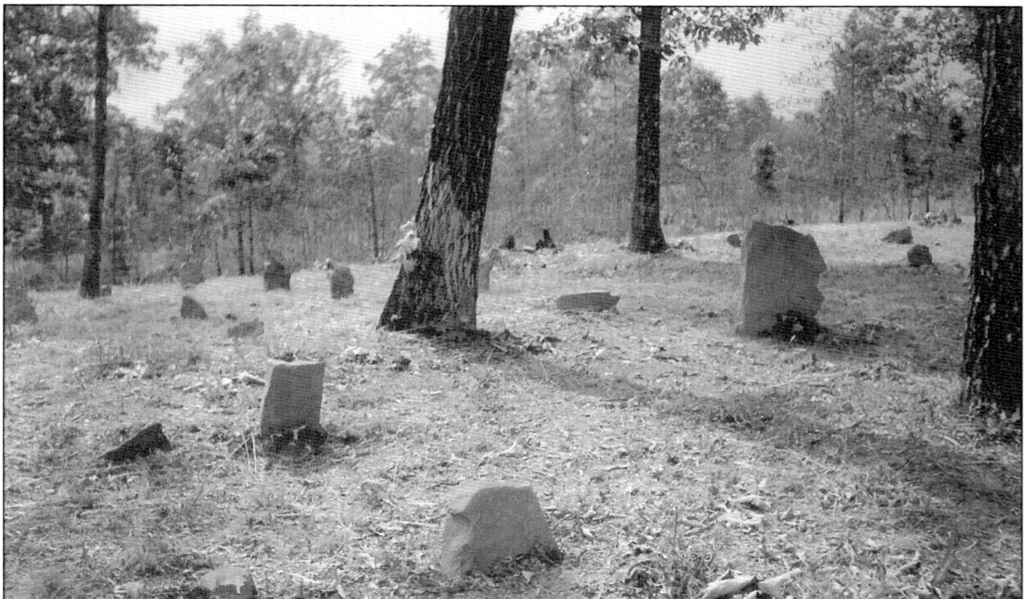

Traditionally, many slave burials took place a night. The coffin, a rough, homemade affair, was placed upon a mule-drawn cart with a line of mourners formed in the rear marching two deep and singing well-known hymns. As they proceeded from the slave quarters to the "lonesome graveyard," the tradition goes that every 15th person would carry an uplifted torch.

For years after the attack by Federal cavalry, Bob Fuller, once a Tannehill Furnace slave, would sit around the Goethite Commissary in the mining area and tell the story of how iron was made. Historian Ethel Armes called him a "Tannehill survivor." It is not known if Fuller lived in the slave quarters near the furnace site, an area that was uncovered by archaeological excavations begun in 2007.

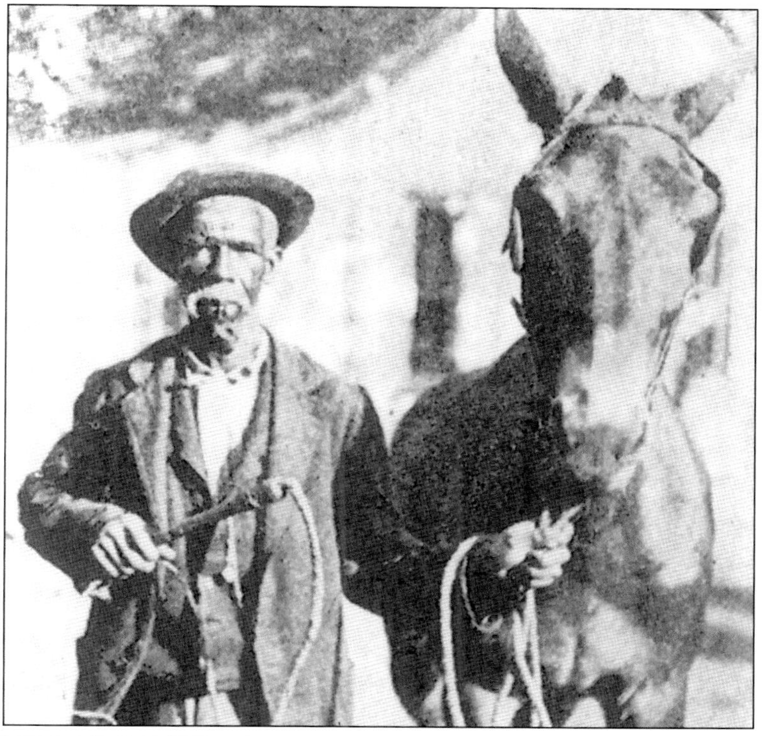

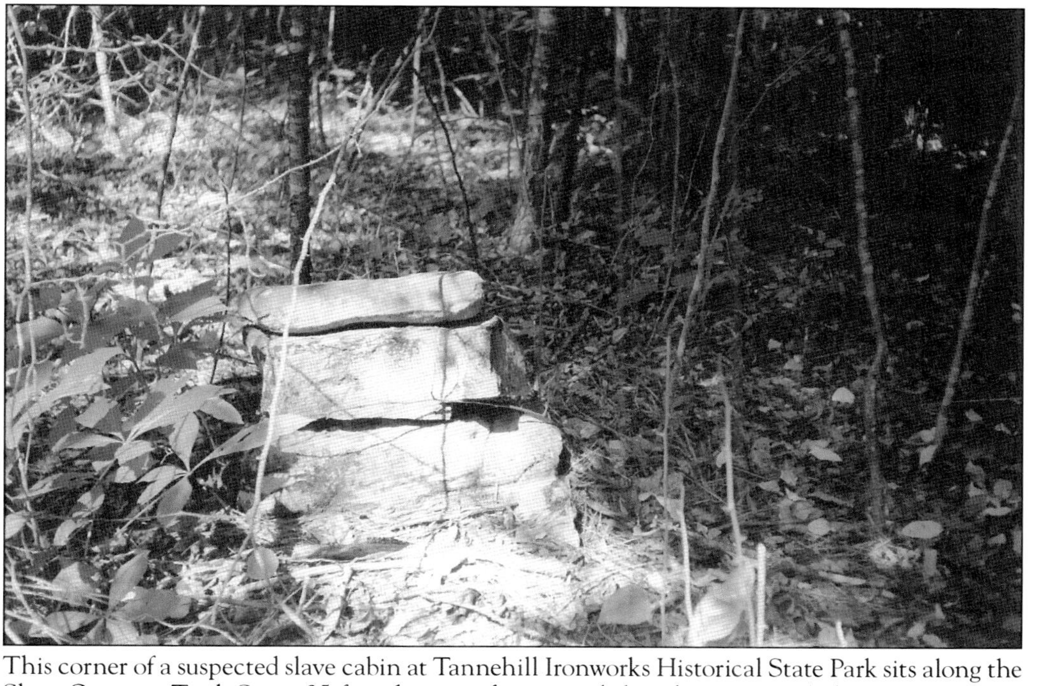

This corner of a suspected slave cabin at Tannehill Ironworks Historical State Park sits along the Slave Quarters Trail. Some 25 foundations of suspected slave houses—including 16 to 20 in a street setting—have been located near the old furnace site. Several hundred slaves once worked in the iron mill.

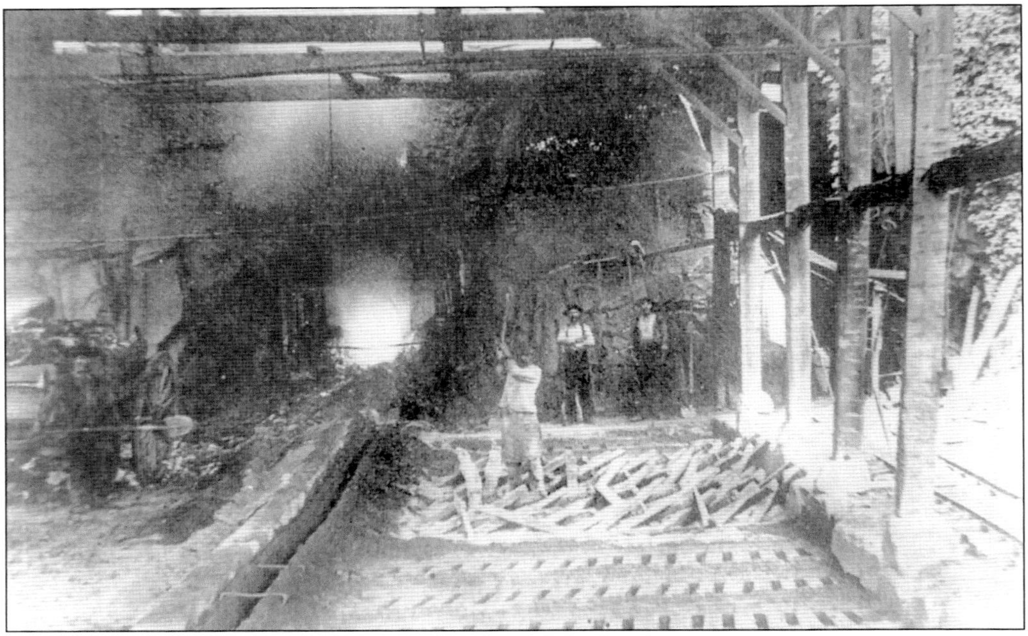

The cast house at the Olive Furnace in Ohio in the 1880s was typical of charcoal blast furnaces around the country, including Tannehill. Pig beds on the right are being broken up into ingots ready to ship to foundries. Located in the Hanging Rock Region in Lawrence County, the furnace built in 1846 operated for 63 years. (Smithsonian Institution.)

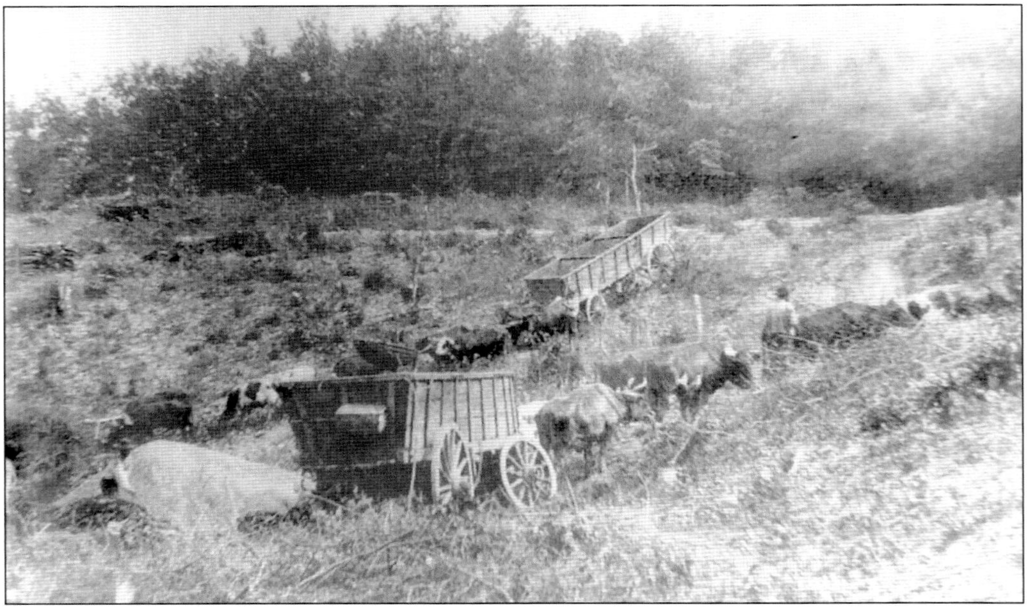

Hundreds of acres of trees were clear-cut around iron furnaces in the 19th century. According to historian Ethel Armes, over 3,400 acres were cleared at Tannehill. This 1880s photograph shows cutting operations needed to make charcoal at the Olive Furnace in Lawrence County, Ohio. Ox-pulled wagons of this type were frequently mentioned in Tannehill records. (Smithsonian Institution.)

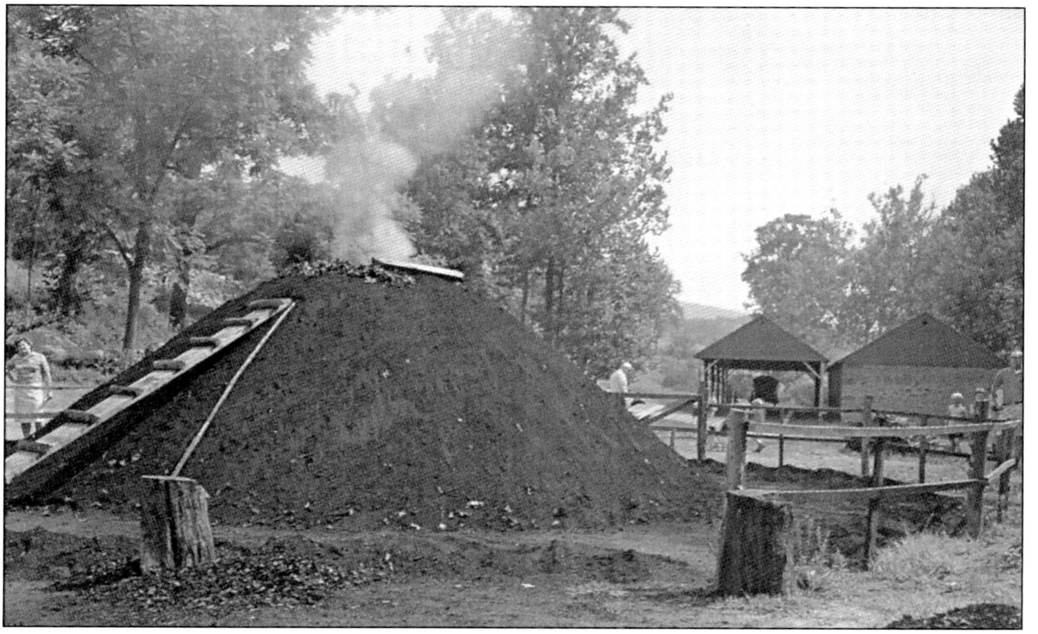

Trees were cut into six-foot sections and stacked into charcoal mounds, which averaged in size from 30 to 40 feet in diameter and were fired for 10 to 14 days. One acre of trees produced only 20 cords, or 800 bushels, of charcoal. Furnaces had voracious appetites. This mound was a charcoal-making exercise at Hopewell Furnace in Pennsylvania. (Walter H. Miller & Company, Williamsburg, Virginia.)

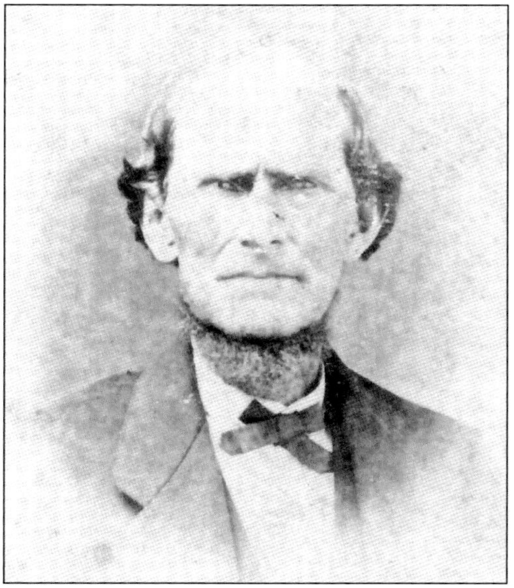

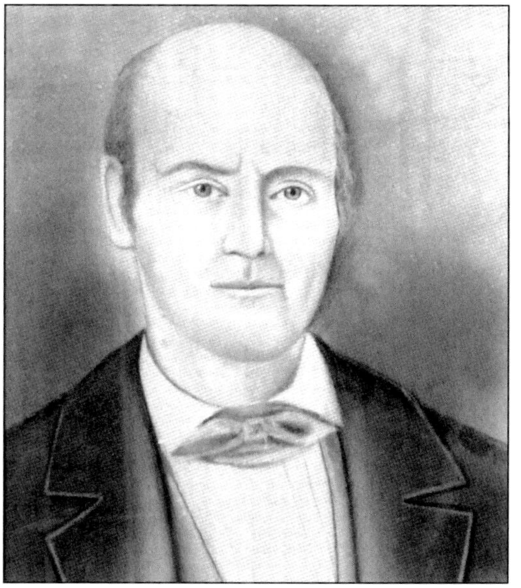

Shown around 1870, John Alexander (left), a wealthy Shelby County planter, bought the Tannehill Furnace property in 1857 from Col. Ninian Tannehill (right). Alexander sold the ironworks to W.L. Sanders in 1862 but regained control after the Civil War. Land grant records show Tannehill, a veteran of the War of 1812, received 40 acres between Hillman's Forge and the present-day tracks of the Norfolk & Southern Railroad in 1836. (Left, Claud E. Thacker, Birmingham, Alabama; right, Ethel Armes.)

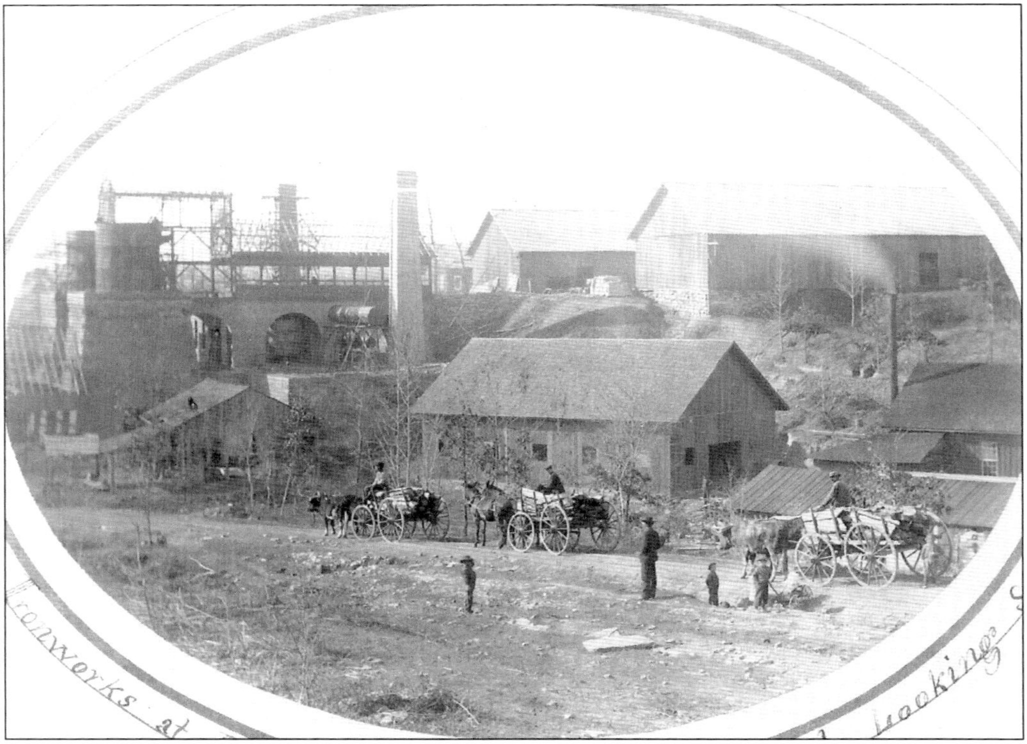

Oxmoor Furnace, built under the supervision of Moses Stroup in 1862, was the first blast furnace in Jefferson County. Stroup left Tannehill at the Confederate government's urging to expand war production. Repaired from war damage and enlarged by the time this image was taken in 1873, it was here in 1876 that an experiment proved good-quality iron could be made from locally produced coke. (Alabama Department of Archives & History.)

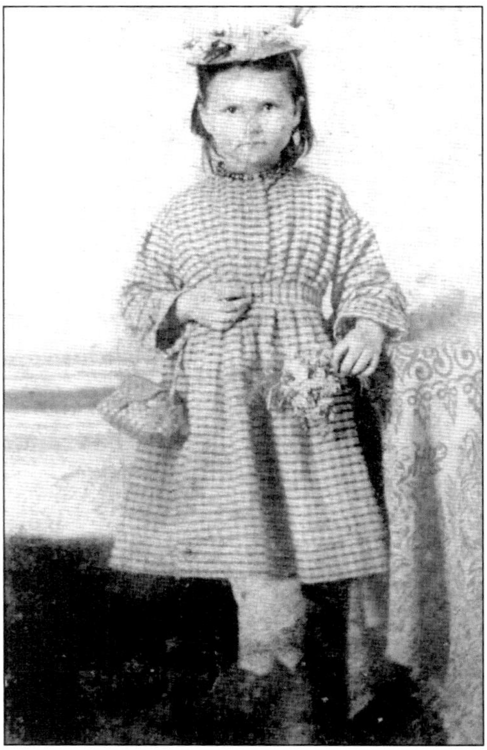

This is a c. 1870 photograph of seven-year-old Capitola Alexander. She was the adopted daughter of the main investor in the Tannehill Ironworks, John Alexander, and his wife, Malinda. Family records indicate that the couple, being childless, adopted her some time after 1863. Since she listed her name as Capitola Black Miskelley after her marriage, there is speculation her father was John Black, a blacksmith working at the Tannehill Furnace during the war. (Claud E. Thacker.)

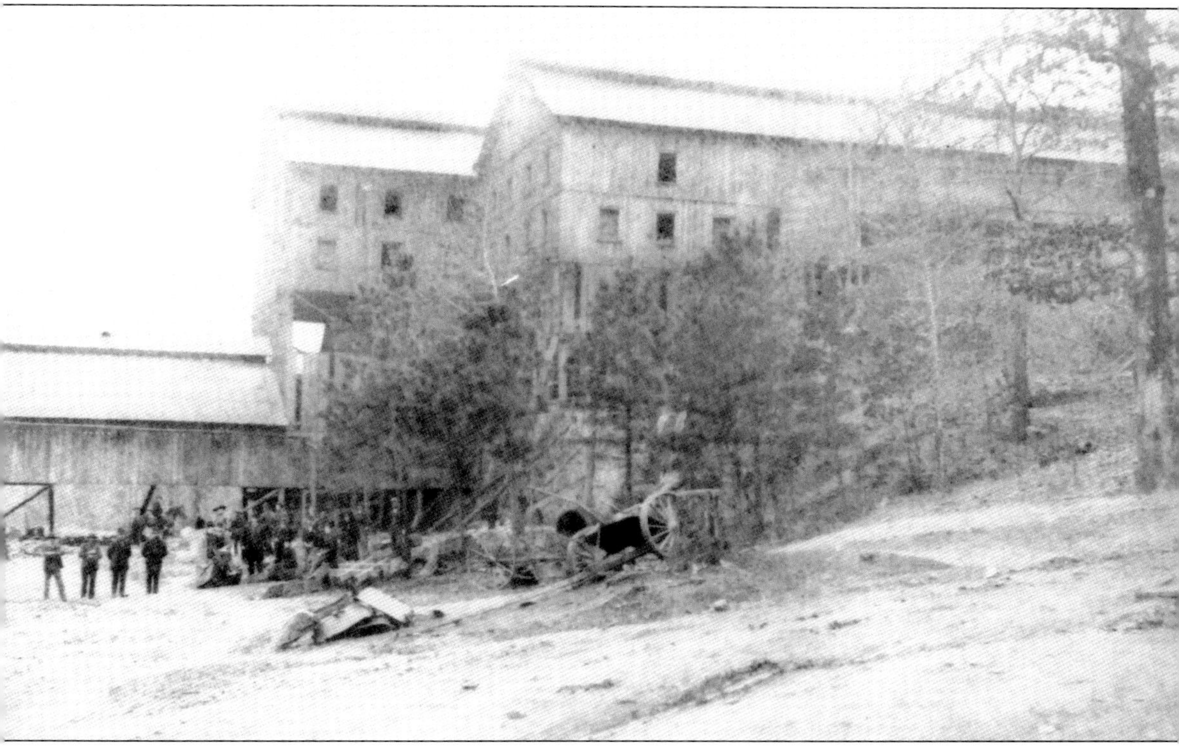

Moses Stroup's first Alabama furnace was the Round Mountain Furnace, near Centre, which he put in blast in 1852 on the site of Milner Forge. At the furnace, Stroup used red ore, its first recorded use in an Alabama furnace. The plant was steam powered. Stroup sold his interest in 1855 and moved to Tannehill to build the furnaces there. (Gadsden Public Library.)

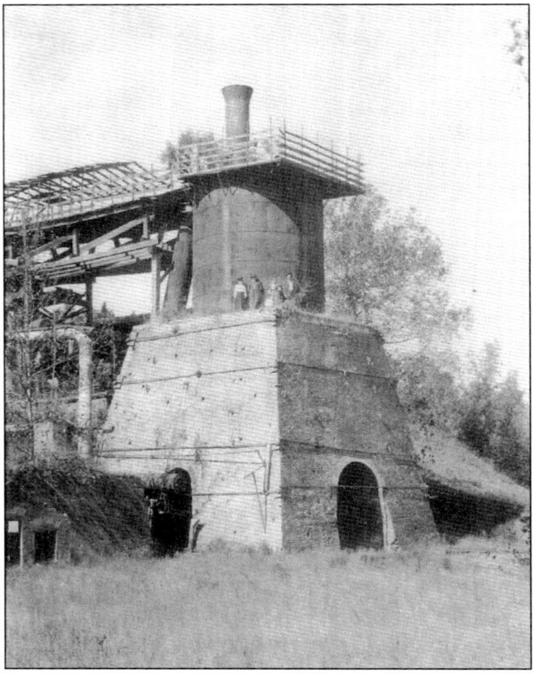

Under the direction of C.C. Huckabee and Jonathan Newton Smith, the Brierfield Ironworks, also known as the Bibb Naval Works, began production of iron for domestic markets in 1862. It was located five miles southwest of Montevallo in Bibb County. When the owners refused to sell all their output for military needs, the CSA War Department, under the command of Maj. W.R. Hunt, took over the works and added a larger second furnace in 1863. This stack became famous for the iron used in the manufacturing of the huge Brooke cannon at the Selma Gun Works. War materials were also processed at a nearby rolling mill, situated along the tracks of the Alabama & Tennessee River Railroad. The furnace and its rolling mill came under attack at the end of the war but were rebuilt. The plant finally closed in 1894. This c. 1910 photograph shows how it looked following closure. (Anna Pfaff, Eugene E. Jones Jr. Collection.)

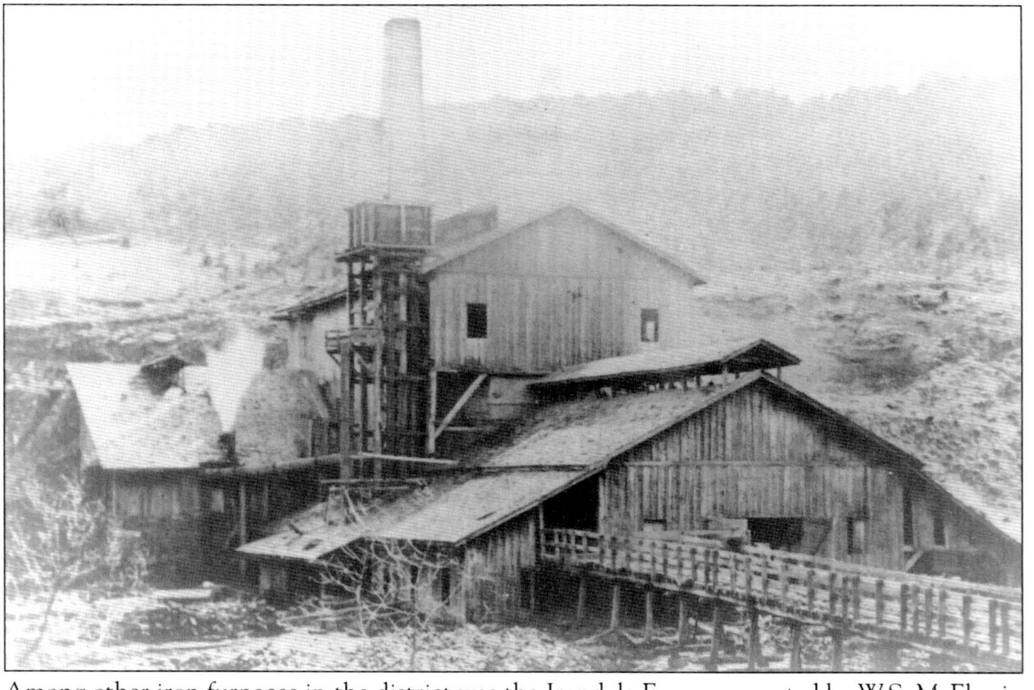

Among other iron furnaces in the district was the Irondale Furnace, erected by W.S. McElwain with Confederate aid in 1863. It was Jefferson County's second blast furnace. Some time during the spring of 1864, the first use of coke as a fuel replacement for charcoal was made here. After the war, it became the first Alabama furnace to go back into production in 1866. (George Gordon Crawford Collection, Samford University Library.)

Three

The Civil War

With iron furnaces falling behind the Federal lines elsewhere in the South, Alabama's role in iron manufacture soared to new heights. It also invited Federal invasion.

During the last two years of the war, Alabama furnaces were turning out 70 percent of the South's entire iron supply, most of it going to the CSA Naval Gun Works at Selma. The fact was not lost on Federal war planners, who sought to end the fighting by cutting off the resupply of iron for Confederate warships, artillery shells, and cannons. To do so, US generals Grant and Thomas proposed a cavalry raid into central Alabama to target coal mines, ironworks, ammunition manufacturers, and other war industries. It would become the largest cavalry raid of the Civil War.

Grant selected Gen. James H. Wilson, a veteran cavalry commander, to lead the raid. Wilson organized and trained approximately 13,480 cavalry troops in northwest Alabama for several months prior to launching the invasion on March 22, 1865. As his forces moved through central Alabama, they met with little resistance from Confederate forces, most of which had been sent eastward during the previous winter to oppose Sherman's March to the Sea. Left to face Wilson's "Yankee Blitzkrieg" was Confederate general Nathan Bedford Forrest, who sent his remaining 5,000-man cavalry to block the advance to Selma.

Next to the Tredegar Ironworks in Richmond, Selma was the largest center for manufacture of war materials in the South. In 1864, it was making two-thirds of all the fixed ammunition for Southern forces and half of the cannons. The facility made the battleships CSS *Tennessee*, *Gaines*, and *Morgan*. It also produced the huge Brooke cannon.

Only the Hale & Murdock Furnace in Lamar County escaped Wilson's torch.

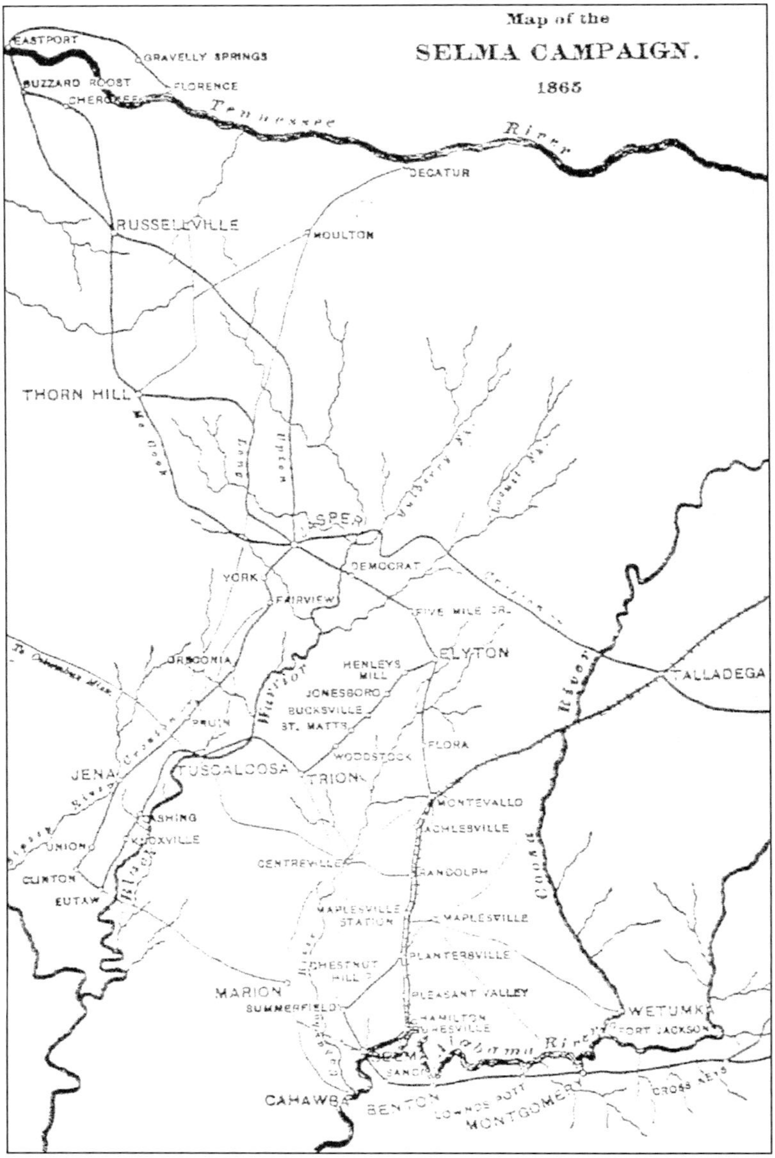

Wilson's Raid into Alabama and Georgia in late March and early April 1865 was the largest cavalry raid of the Civil War, involving over 13,000 troops. The main targets of the raid were the 12 Alabama blast furnaces and the Selma Arsenal and Gun Works. This battle map of the Selma campaign from the Ohio Commandry, the 1888 Military Order of the Loyal Legion, misidentifies McMath's (post office) near Tannehill as "St. Matts." Wilson's invading force consisted of three columns led by Gens. Eli Long, Edward M. McCook, and Emory Upton. Under tight military security, the columns split to mask Wilson's intentions and to confuse the enemy as they moved from Florence to Elyton. At the later place, 1,500 cavalry split from McCook's division under Brig. Gen. John T. Croxton to attack the ironworks at Tannehill and the city of Tuscaloosa, the site of the University of Alabama, which was considered a military school. Even Nathan Bedford Forrest did not know until late that Wilson's main target was Selma. (Lewis M. Hosea, *Sketches of War History: 1861–1865.*)

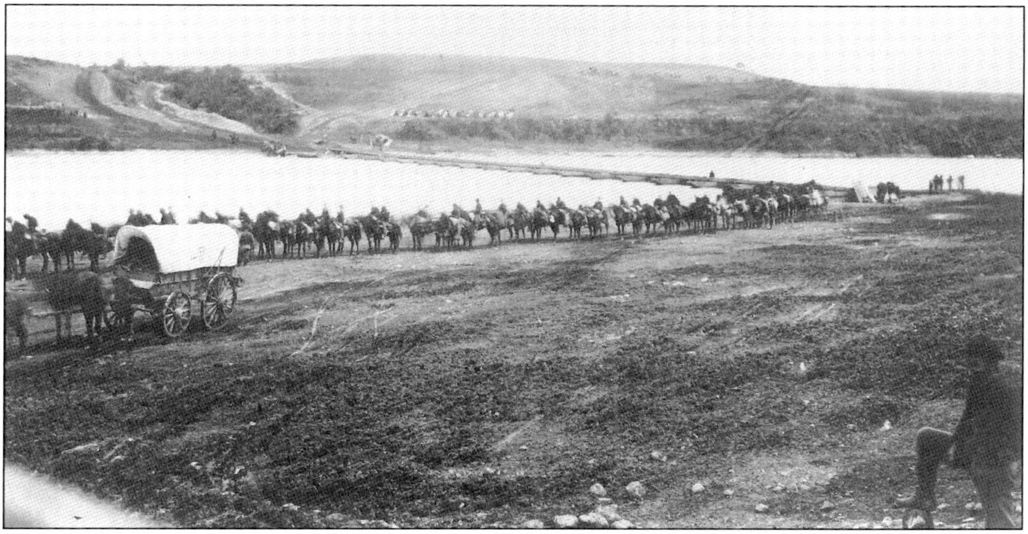

This is a view of Federal cavalry on the move crossing the Rappahannock River near Fredericksburg, Virginia, early in the war. This scene is similar to Wilson's crossing of several Alabama rivers, including the Warrior, in 1865. Wilson's invasion force began its 525-mile journey in Lauderdale County on March 22 and, after destroying the state's iron industry and the Selma Arsenal, ended up at Macon, Georgia, on April 20. (US Army Military History Institute.)

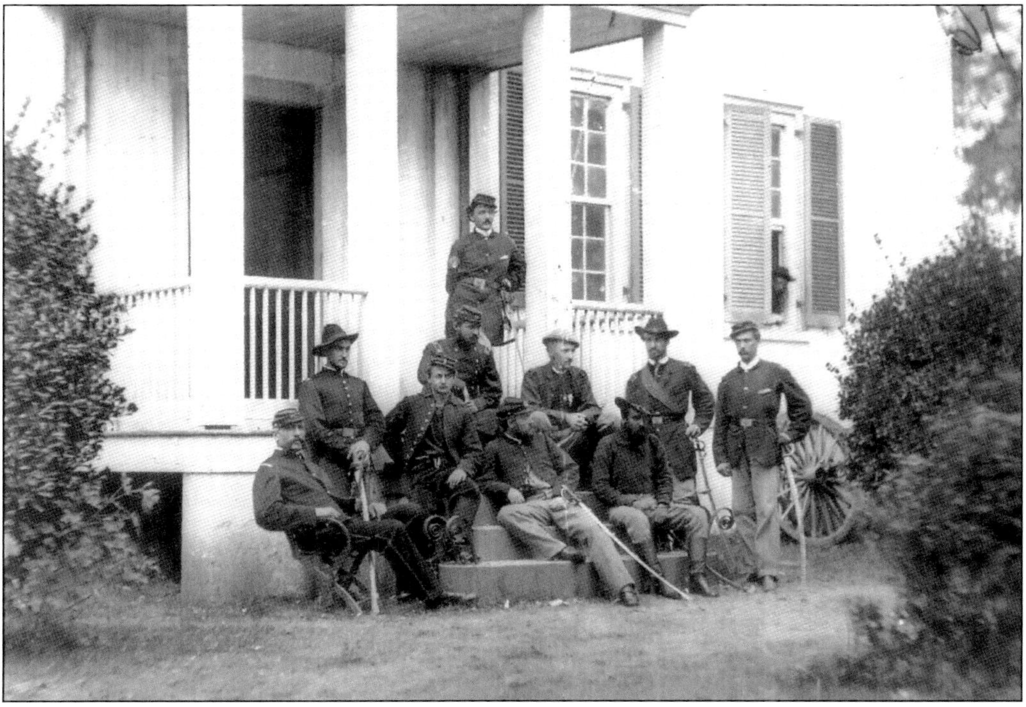

General Wilson (center) and his staff pose in this Brady photograph taken just before the Reams Station Raid on August 25, 1864, a part of the campaign for Petersburg, Virginia. Six months later, he would be in Alabama, where soldiers under his command camped from Gravelly Springs to Waterloo, just north of the Tennessee River. (US Army Military History Institute.)

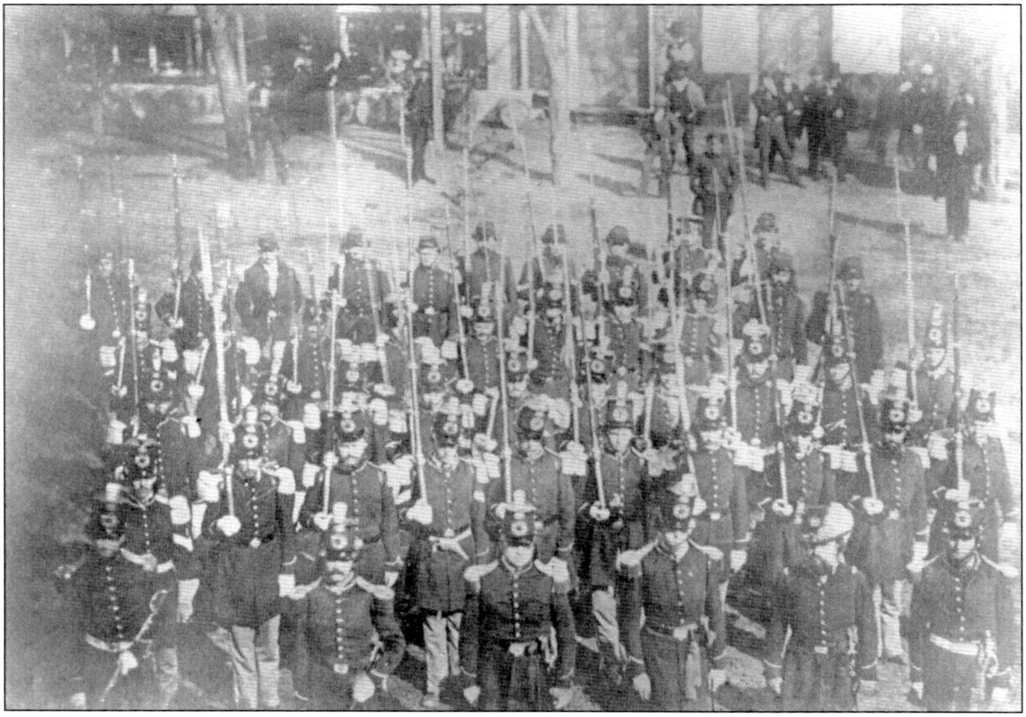

The Independent Blues, an early Confederate unit, posed for this photograph on a downtown Selma street in 1861. This unit became Company D, 8th Alabama Infantry, assigned to the Army of Northern Virginia. The storefronts in the rear may have been similar to Cavett & Sanders Mercantile on Water Street. Sanders bought the Tannehill Furnaces a year later. (Alabama Department of Archives & History.)

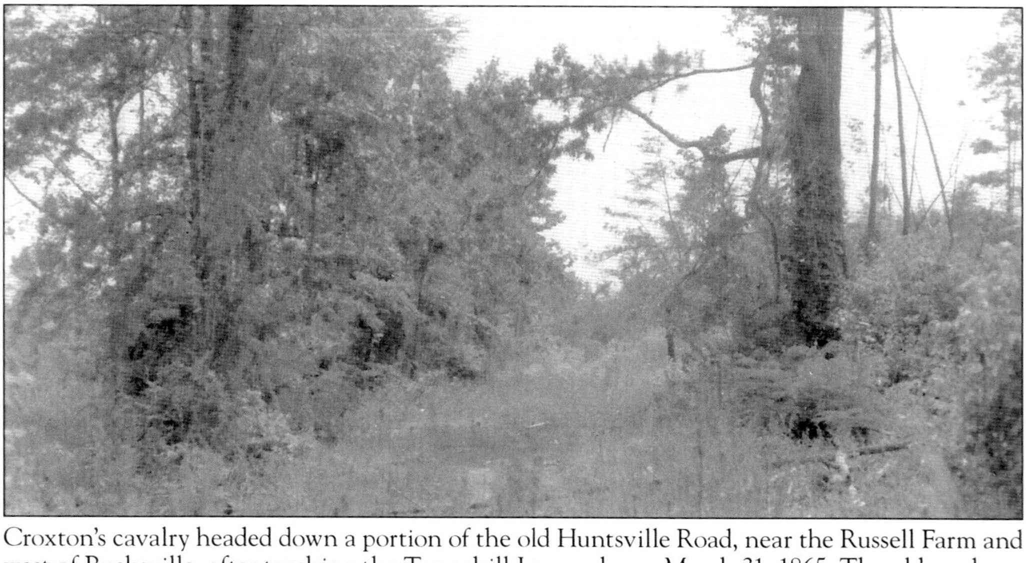

Croxton's cavalry headed down a portion of the old Huntsville Road, near the Russell Farm and west of Bucksville, after torching the Tannehill Ironworks on March 31, 1865. The old roadway, still visible in places, ran near the big two-story house of the postmaster, Elisha McMath, and a nearby spring. (Jim Bennett, Birmingham, Alabama.)

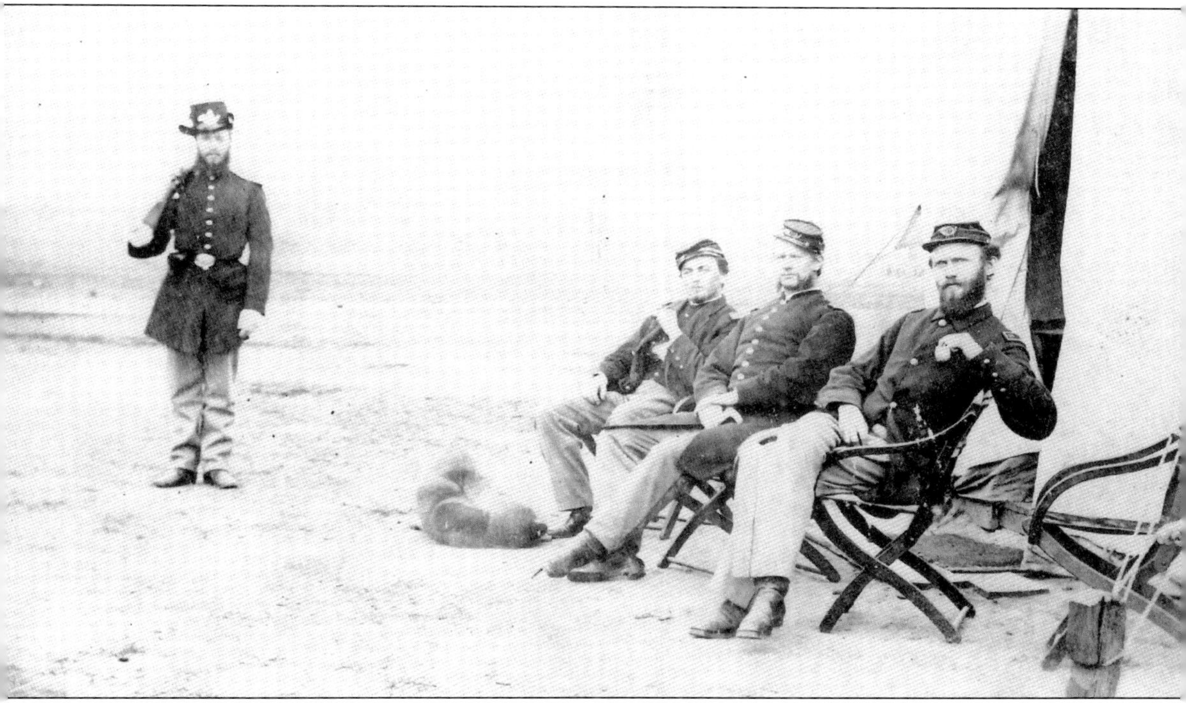

The Tannehill Furnace raid was led by Capt. William A. Sutherland (seated center), who was commanding three companies of the 8th Iowa Cavalry that were dispatched from Croxton's main column and headed toward Bucksville. Sutherland is shown in this photograph as a member of the 17th US Infantry, stationed in Hempstead, Texas, in 1866. Seated with him are, from left to right, 2nd Lt. T.G. Troxel and Capt. George Lancaster. Sutherland, who served as General Croxton's adjutant general during the Civil War, led several other Alabama raids, including one on April 5 where he burned the Pickens County Courthouse at Carrollton. After being discharged from the army, Sutherland died in Texas of yellow fever in 1867 and is buried in Houston. He never returned to his mother and siblings in Steubenville, Ohio, where he joined the 2nd Ohio Volunteer Infantry at the start of the Civil War. (Ohio Historical Society.)

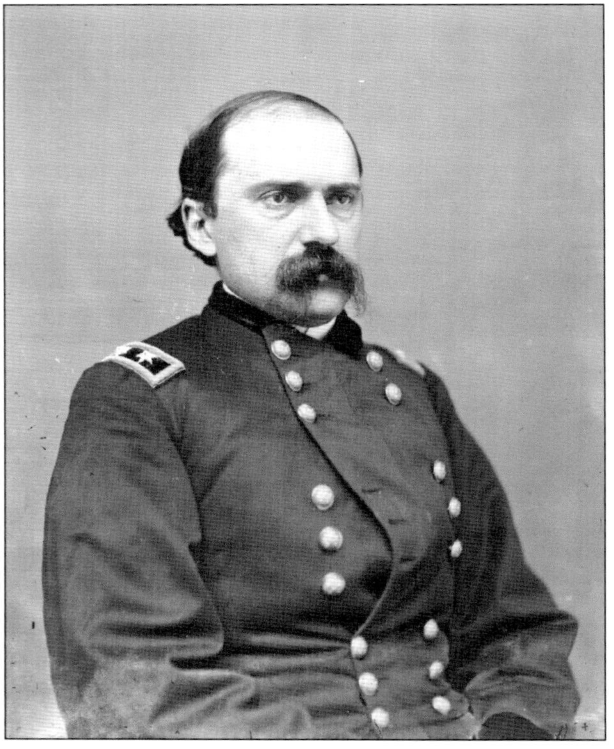

Gen. Edward M. McCook, who commanded the 1st Division in Wilson's Raid through Alabama and Georgia in 1865, was a member of the famed "Fighting McCook" family of Ohio. Four of his brothers and 10 of his first cousins served as officers, with six of the family members becoming generals before the end of the war. He later became territorial governor of Colorado. (Library of Congress Prints and Photographs Division.)

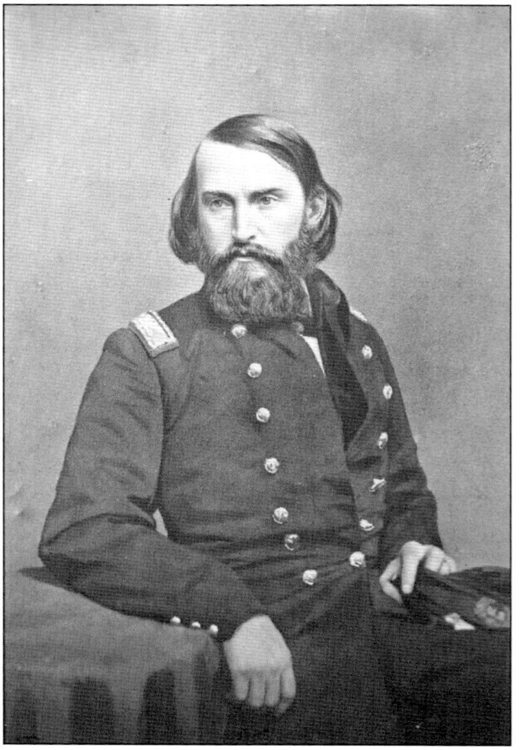

On April 4, 1865, Gen. John T. Croxton, shown around 1862, captured Tuscaloosa, strengthening the Union army's grip on central Alabama. In the process, he burned most of the University of Alabama's buildings, as well as the town's industries and warehouses. A part of this force attacked the Tannehill Ironworks five days earlier. Croxton unsuccessfully attempted to buy the Oxford and Shelby ironworks and the Selma Gun Foundry after the war. (Library of Congress.)

Near the close of the war, Gen. William Wert Adams served with Gen. Nathan Bedford Forrest in opposition to Wilson's Raid in Alabama. Near Pickensville, he blocked Gen. John T. Croxton's attempt to rejoin Wilson at Selma after his attack on Tuscaloosa. In pursuit, Adams's brigade engaged Croxton in a furious firefight at Romulus. After the war, he served as postmaster in Jackson, Mississippi. (Alabama Department of Archives & History.)

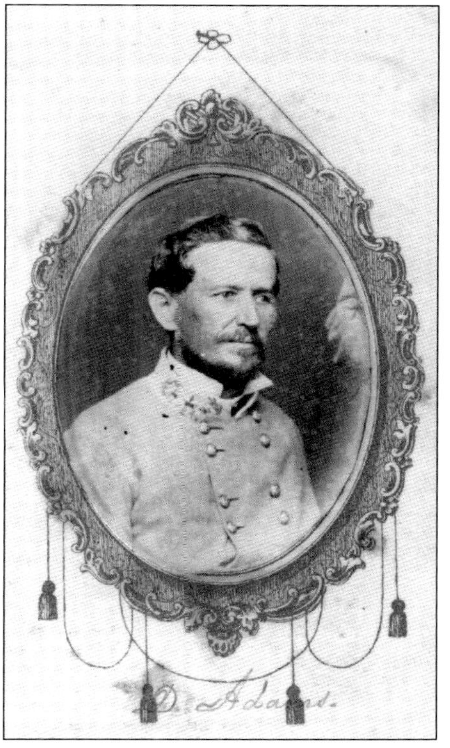

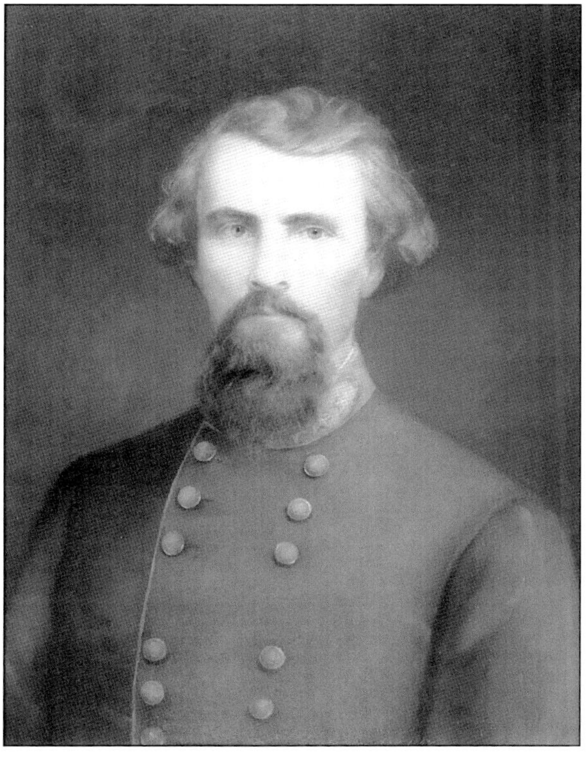

A Confederate cavalry commander, Lt. Gen. Nathan Bedford Forrest was one of the war's most unusual figures. Known as the "wizard of the saddle," his very presence struck fear in enemy ranks. Forrest reputedly had a total of 30 horses shot out from under him and killed 31 people. In 1865, he attempted, without success, to defend Alabama against Wilson's Raid at Selma. (Alabama Department of Archives & History.)

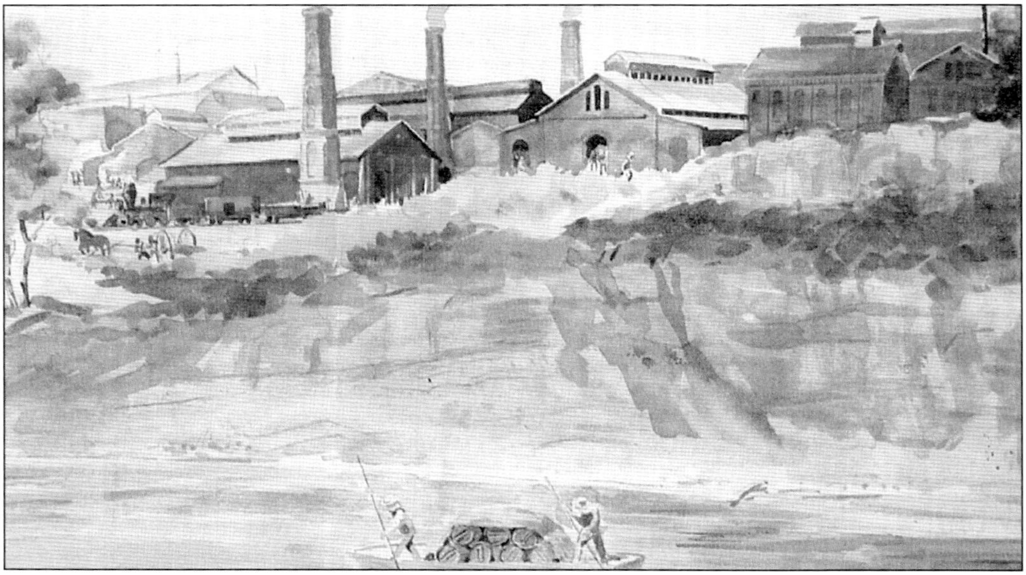

The Selma Arsenal and Naval Gun Works at its height was surpassed only by the Tredegar Ironworks in Richmond, Virginia, as a producer of Confederate war materials. These were the only two foundries in the South that could cast the huge Brooke cannon, the Confederacy's largest artillery piece. The arsenal employed over 6,000 workers making everything from shovels to battleships. (Frances Lanier portrait, Old Depot Museum, Selma, Alabama.)

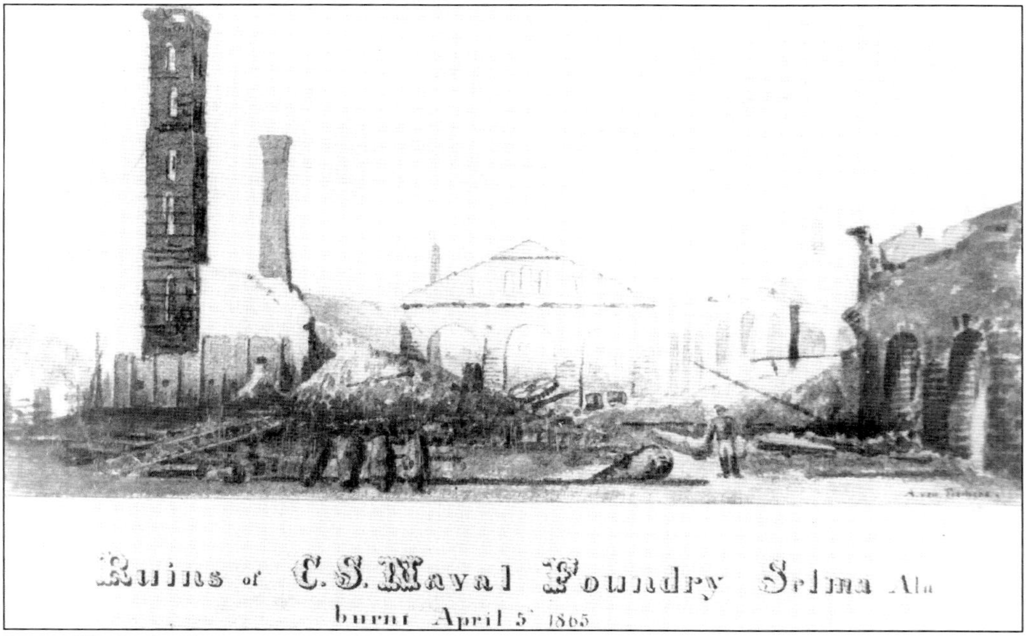

Burned by Federal troops on April 5, 1865, here are war ruins of the Confederate naval foundry at Selma. After breaking through Confederate defenses held by 5,000 troops under Forrest, General Wilson took Selma on April 2. Forrest narrowly escaped capture. The Federal force went about burning buildings and destroying arms, some of which were dumped in the Alabama River. (Alabama Department of Archives & History.)

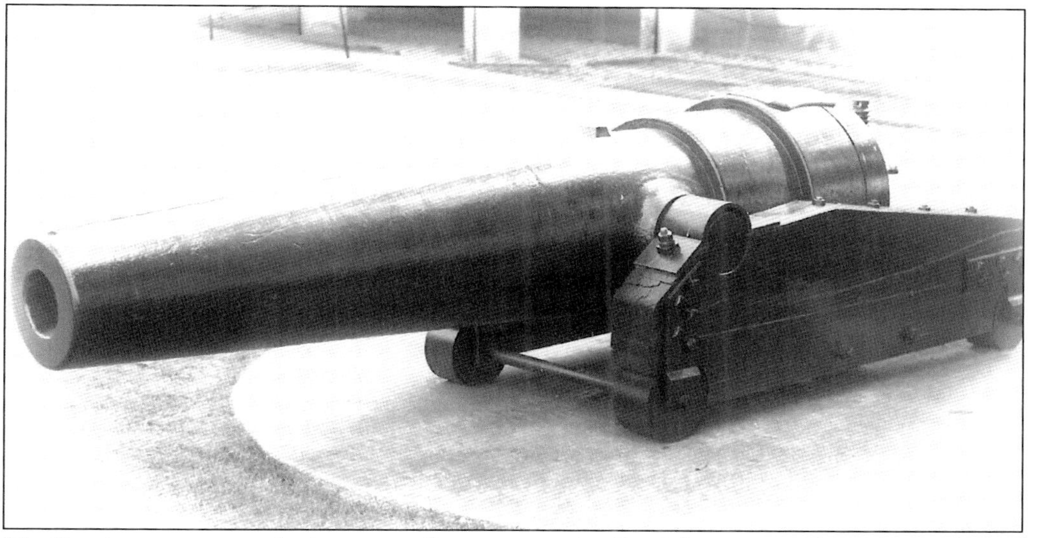

The Brooke cannon was the largest artillery piece in the Confederate arsenal. Made at Selma of Brierfield Furnace iron, the huge gun could hurl a 230-pound projectile 2,200 yards. This Brooke was salvaged from the gunboat CSS *Chattahoochie*. The Brooke rifle is similar to the Federal Parrott in appearance. Guns manufactured at Selma bear the foundry's imprint of "S," and those from Tredegar had the imprint "TF." (Confederate Naval Museum, Columbus, Georgia.)

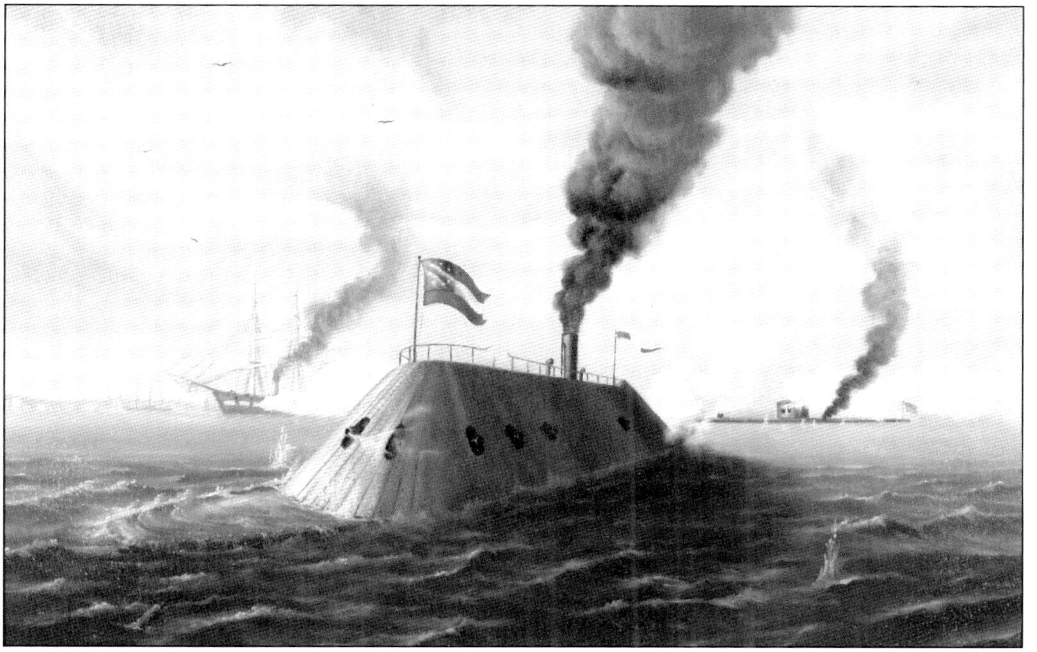

The Naval Gun Works at Selma made several famous warships, including the CSS *Tennessee*, *Gaines*, and *Morgan*. Ironclads like these were steam-propelled warships protected by iron or steel armor plates. After the clash of the Federal *Monitor* and the Confederate *Merrimack* (*Virginia*), depicted here, on March 9, 1862, the era of wooden ships was over. Iron from Alabama furnaces was used in armor plating for the Selma-made gunboats. (Raymond Bayless illustration, US Naval Historical Center photograph.)

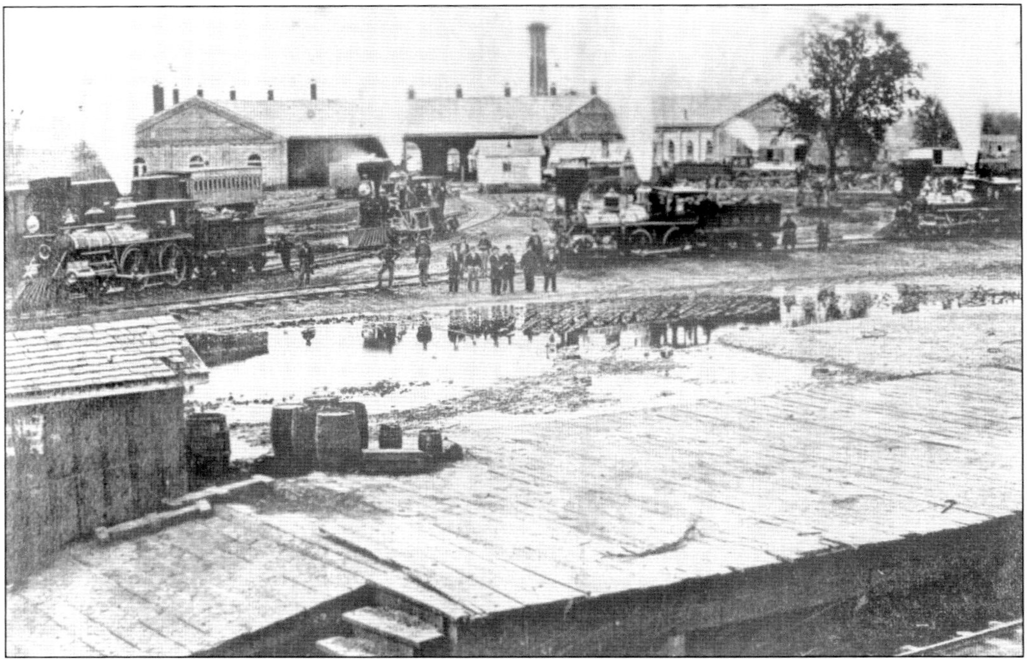

US military locomotives occupy the train yard at the roundhouse and shops of the Memphis & Charleston Railroad shortly after the Federal occupation of Huntsville in 1864. No photographs have been found of the Alabama & Tennessee River Railroad Station at Montevallo, but it may have had a similar look. (Huntsville–Madison County Public Library.)

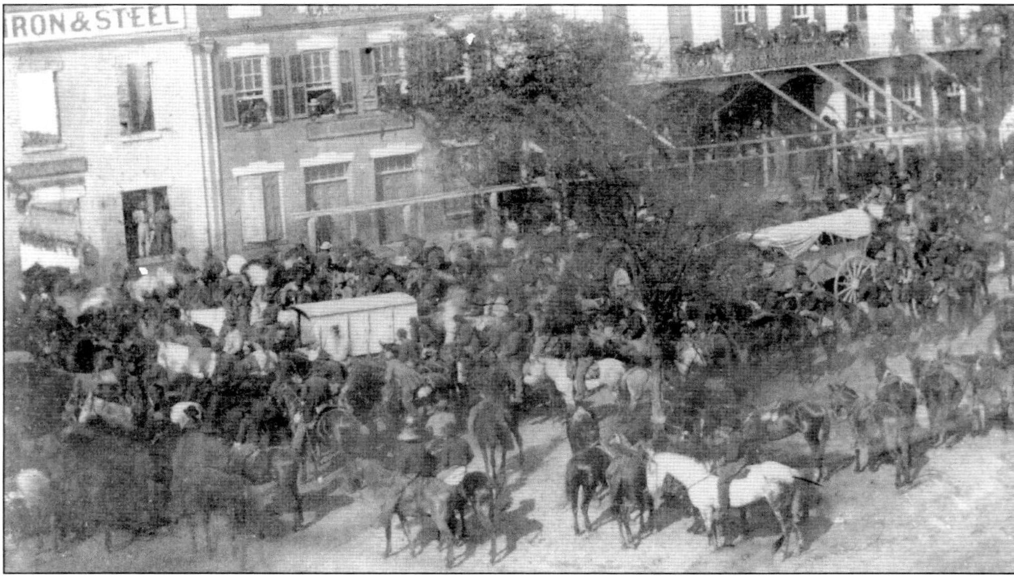

Photographs of Wilson's Raid are also hard to find. This photograph, however, does show some of Wilson's troops soon after the capture of Jefferson Davis by the 4th Michigan Cavalry on May 10, 1865, in Irwinsville, Georgia. Here, Union troops surround an ambulance carrying the former Confederate president and his family through Macon en route to Washington, DC. (A.J. Riddle photograph, Bettmann Archive.)

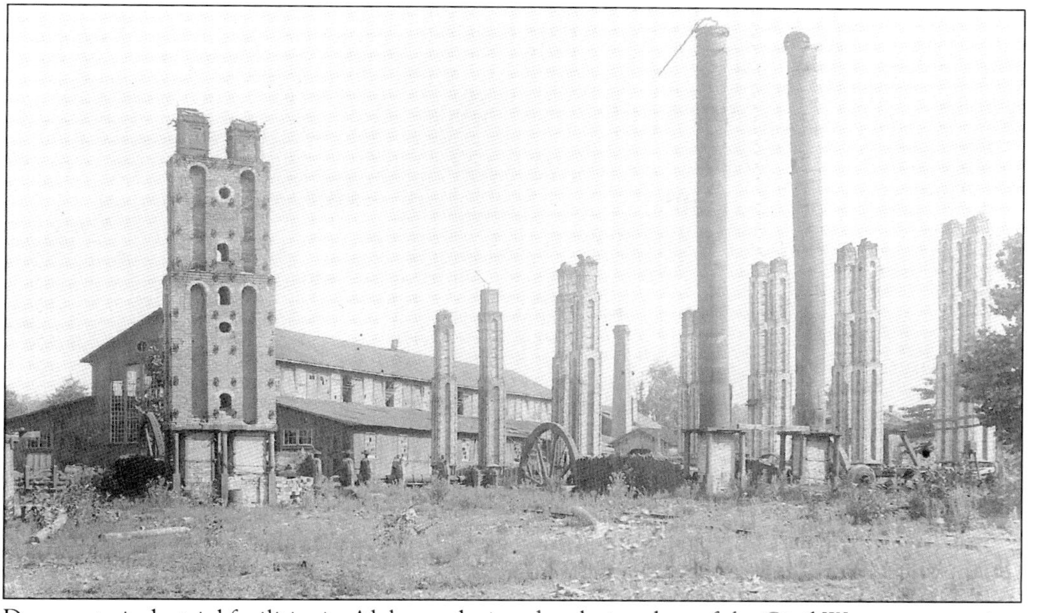

Damage to industrial facilities in Alabama during the closing days of the Civil War was extensive. Here are the ruins of the Brierfield Rolling Mill, standing forlorn and silent around 1900 near the tracks of the Alabama & Tennessee River Railroad in Bibb County. Although rebuilt after the war, it fell into disrepair, and both the mill and furnace closed in 1894. (Anna Pfaff photograph, Eugene E. Jones Jr. Collection.)

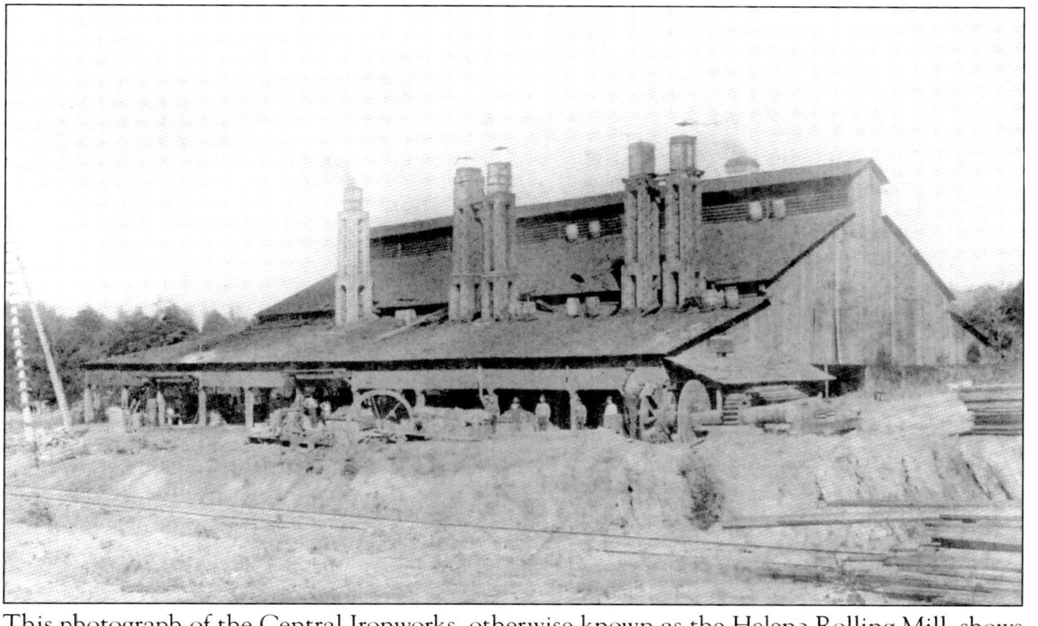

This photograph of the Central Ironworks, otherwise known as the Helena Rolling Mill, shows it in operation in 1875. A war casualty, it had just been completed when attacked by Federal cavalry on March 30, 1865. After lying dormant for seven years, the property was bought at a tax sale, and a new company, making repairs, put it back in operation in 1873. (Ken Penhale, Helena, Alabama.)

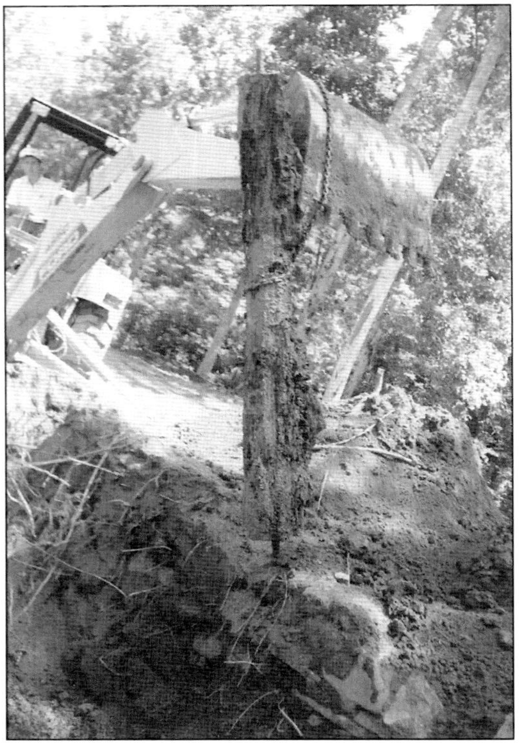

A backhoe operator lifts a piece of charred post from the blower house of the Tannehill Ironworks in 1990 as restoration gets underway on the facility. The building housed waterwheels that were needed for air blast on Furnace No. 1. The waterwheel was also used to operate a machine shop or refinery forge. The building was burned by Federal cavalry on March 31, 1865, during Sutherland's raid on the iron-making facility.

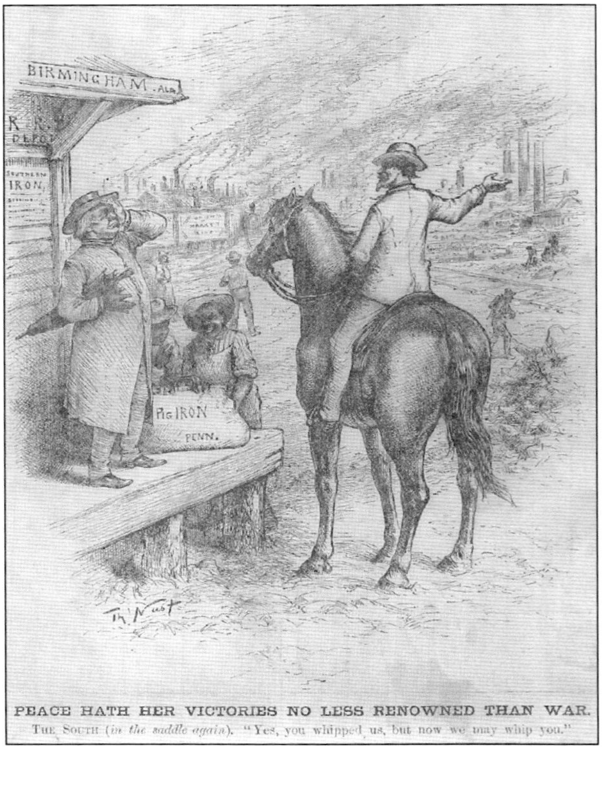

PEACE HATH HER VICTORIES NO LESS RENOWNED THAN WAR.
THE SOUTH (*in the saddle again*). "Yes, you whipped us, but now we may whip you."

Although their industry lay in shambles following the Civil War, Alabama iron makers gradually made a comeback beginning with the Irondale Furnace in 1866. All but six of the old Alabama blast furnaces were rebuilt. This Thomas Nast cartoon published in *Harper's Weekly* on March 26, 1877, was a reminder to Pennsylvania ironmasters that even though the North won the war a battle on another front was just beginning. (*Harper's Weekly*, 1877.)

Four

Mining and Milling

After several attempts to rebuild the furnaces at Tannehill fell through following the Civil War, the Pioneer Mining & Manufacturing Company purchased the Tannehill ore mining property in 1868 with an eye toward building new furnaces in Jones Valley. In acquiring the ore fields, the Pioneer Company, a Pennsylvania enterprise led by David Thomas, received some 2,615 acres from John Alexander.

Milling operations at Tannehill involving both corn and wheat also resumed when Hall's Mill was built on Mill Creek in 1867, replacing the Tannehill gristmill burned during the war.

The Tannehill ore beds were extensive. Brown ore (limonite), near Bucksville, was reported to be 100 feet thick in places. One of the first occurrences of strip mining in Alabama has been identified here. Red ore (hematite), like that found on Red Mountain in Birmingham, is more frequently mined by sinking deep sloping shafts.

Several iron and steel companies mined the Tannehill ore sites throughout the late 1800s and early 1900s, including the Central Iron & Coal Company; Tennessee Coal, Iron & Railroad Company; Republic Iron & Steel Company; and the Woodward Iron Company. The Shook and Fletcher Supply Company was the last to extract ore from the field, ending its activities in the 1960s. Old mining sites can be seen just off the Tannehill Parkway, the Waterton Lakes subdivision, and Old Village Road. Ponds were the result of using steam shovels mounted on rails to mine huge quantities of the ore. When an area would mine out, operators would simply move their rails and shovels to the next nearby opportunity.

Mines with names like Giles, McMath, Greeley, Standiford, Martaban, Goethite, Hillman, Reno, Houston, and Woodstock sent thousands of tons of ore to local steel mills.

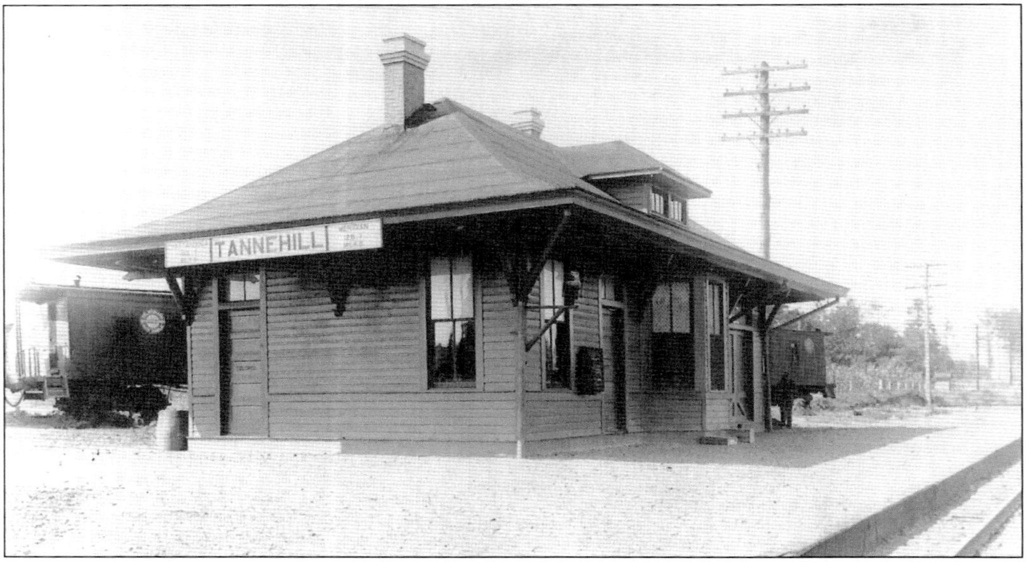

Once a busy stopover for iron ore miners along the Alabama Great Southern Railroad, this is how the Tannehill Station appeared in 1915 before being torn down. The old Tannehill House sat on a high hill behind it. A sign on the station informed travelers they were 166 miles from Chattanooga and 120 from Meridian. (Southern Railroad Archives.)

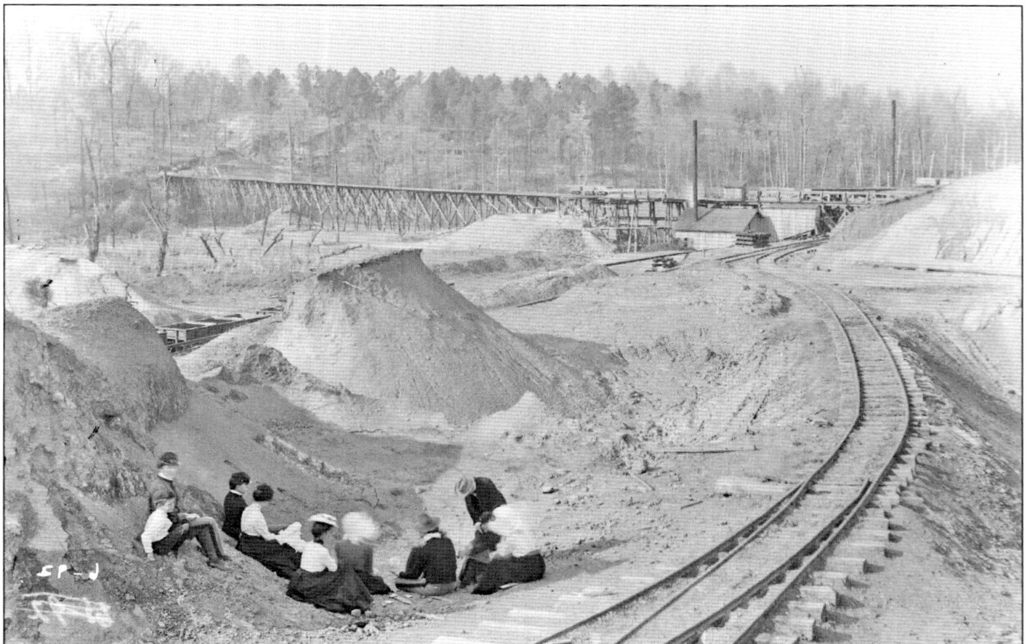

This group of men, women, and children look a bit overdressed on a day trip to the Tannehill Ore Mines around 1900–1919. The activity apparently was common, as mining operations drew great interest from worker families, investors, and mill managers. Exposed columns of Conasauga limestone left from ore extraction rise along the mineral line train tracks. (Eugene A. Smith Collection, W.S. Hoole Special Collections Library, University of Alabama.)

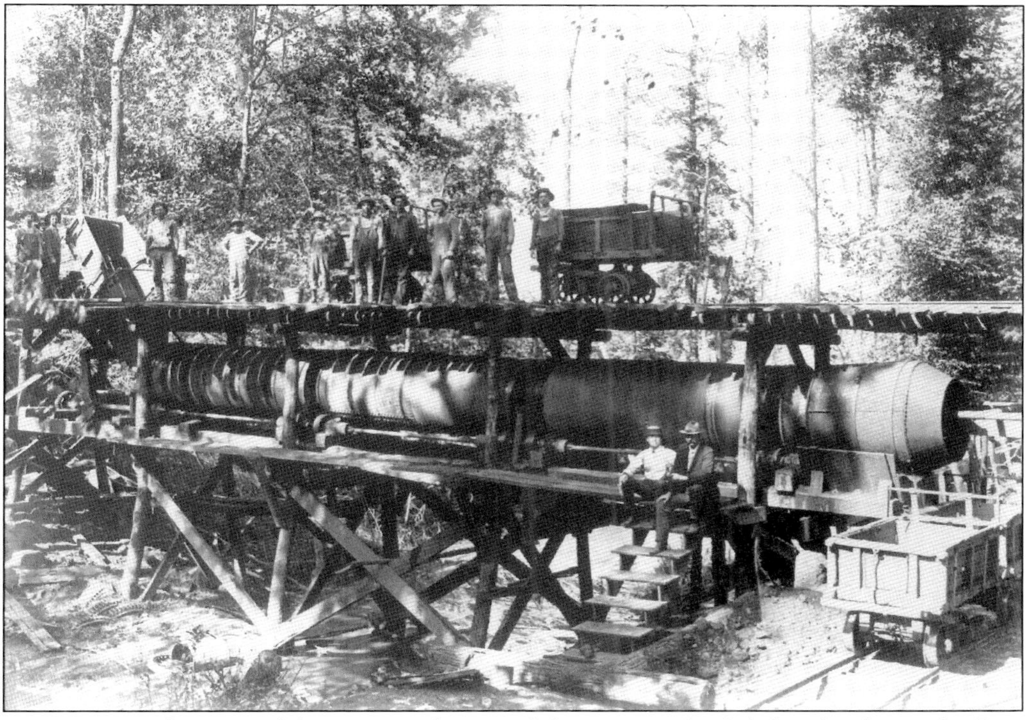

Dump cars in this turn-of-the-century photograph have just unloaded their burden at the Burr ore washer, which was located at the Goethite Mine at Tannehill and operated by the Pioneer Mining & Manufacturing Company. With the waste water finding its way back into Roupes Creek, the waterway became known as Mud Creek. (Birmingham Public Library.)

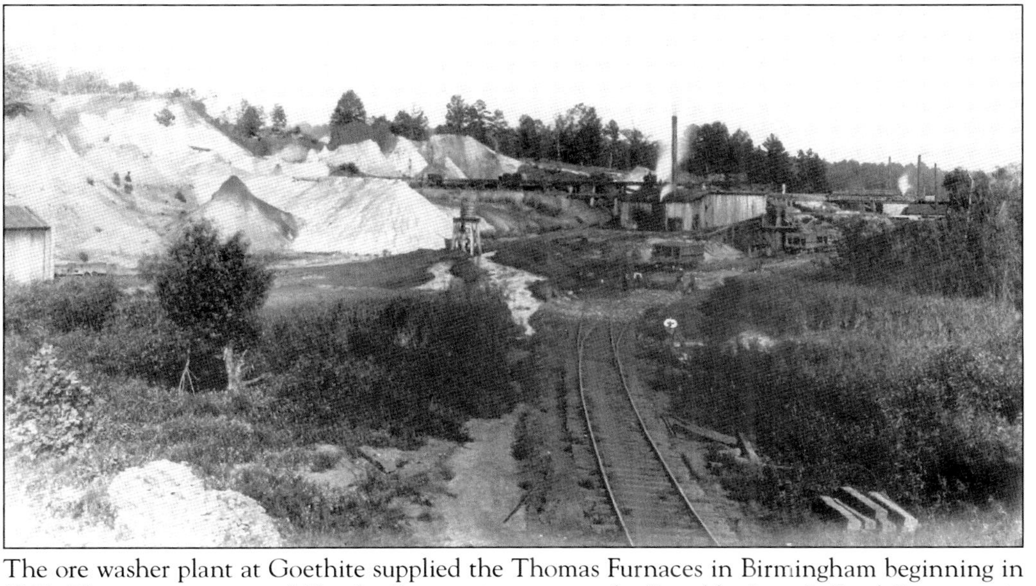

The ore washer plant at Goethite supplied the Thomas Furnaces in Birmingham beginning in 1888. This property was sold by the Pioneer interests to the Republic Iron & Steel Company in 1889. The hills of residual limestone testify to the heavy extent of brown ore mining in the area. (Birmingham Public Library.)

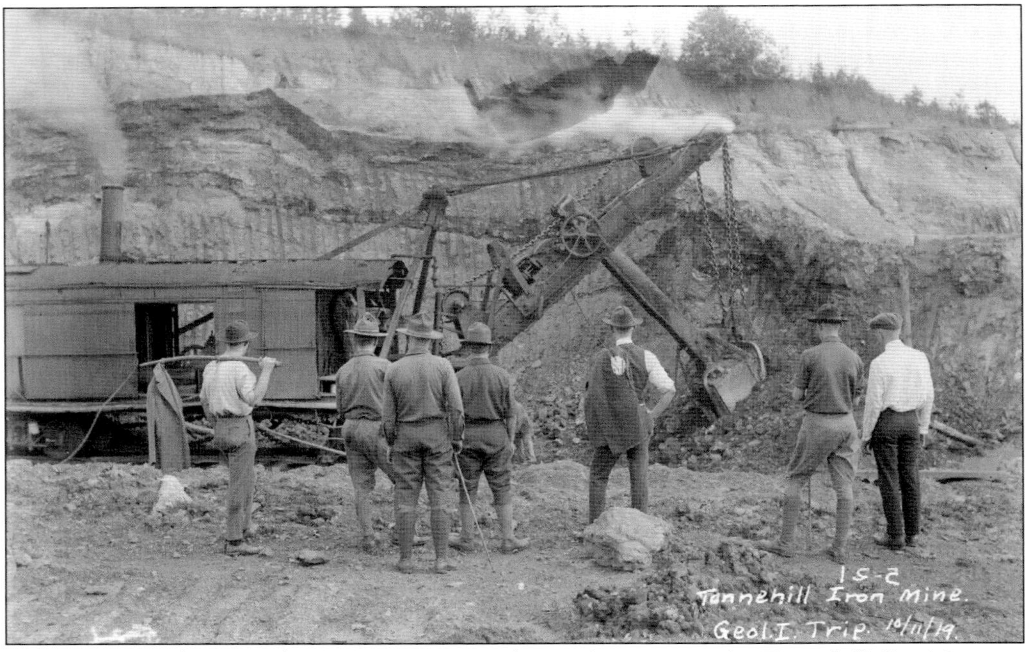

Future state geologist Walter B. Jones was on this geology trip to the Tannehill Ore Mines in 1919 to view the work of a steam shovel excavating brown ore. Jones, who became assistant state geologist in 1925, excavated the Moundville Archaelogical Site and established it as an important Native American burial ground and mound location. (Eugene A. Smith Collection, W.S. Hoole Special Collections Library, University of Alabama.)

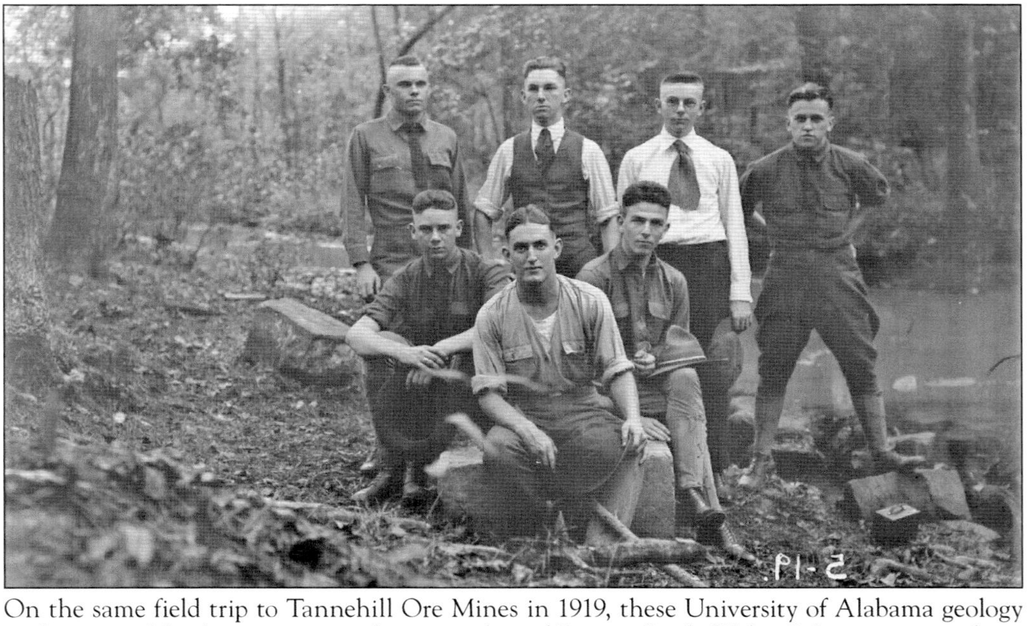

On the same field trip to Tannehill Ore Mines in 1919, these University of Alabama geology students posed for the camera near the ironworks and Roupes Creek. Walter B. Jones was a graduate student at the time and was the photographer. (Eugene A. Smith Collection, W.S. Hoole Special Collections Library, University of Alabama.)

Pictured around 1925 at their home near the gristmill are John Wesley Hall and wife Mary Elizabeth. The Reconstruction-era mill and its rock dam were built in 1867 by Ninian Tannehill, who turned the operation over to Hall to run. In 1870, the mill is recorded having ground 3,720 bushels of corn and 800 bushels of wheat. It closed in 1931 after being damaged by a tornado. (Robert Hall, Bessemer, Alabama.)

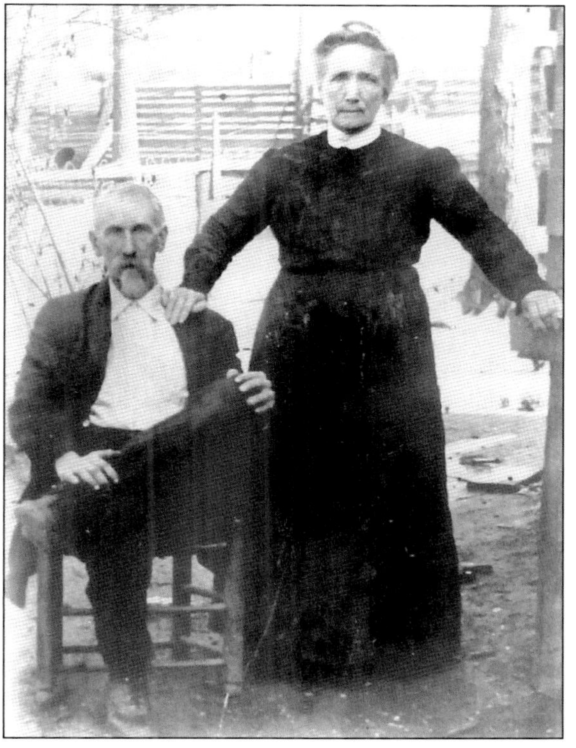

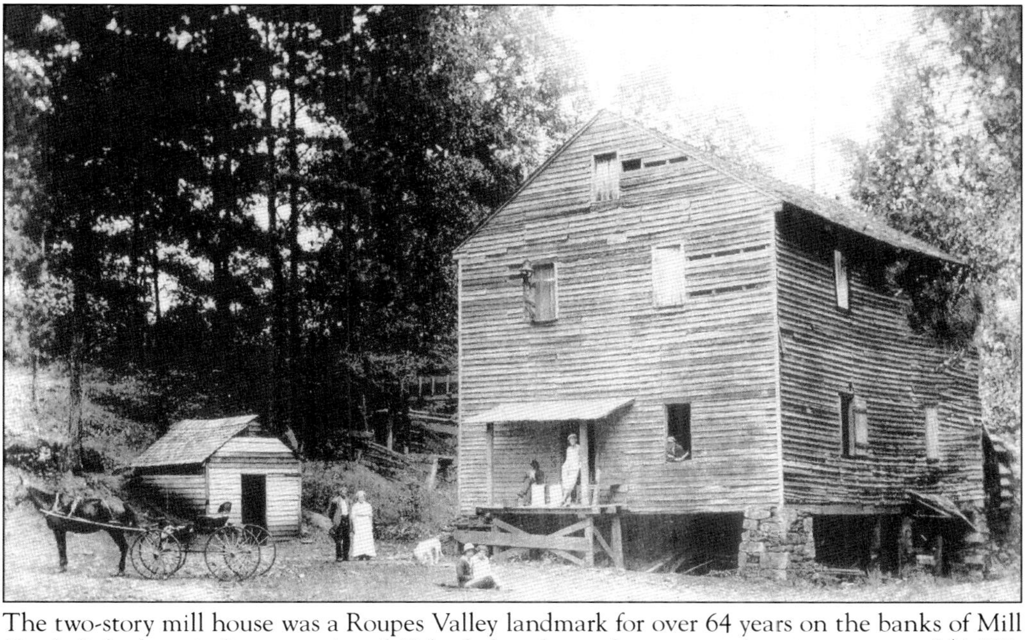

The two-story mill house was a Roupes Valley landmark for over 64 years on the banks of Mill Creek. It had a grinding capacity of 60 bushels a day and represented an investment of $1,500. The operation ground both corn and wheat into meal and flour. The mill also was used as a cotton gin. Possibly used to signal that an order was ready, a bell is located on the second floor. (Neal Countryman, McCalla, Alabama.)

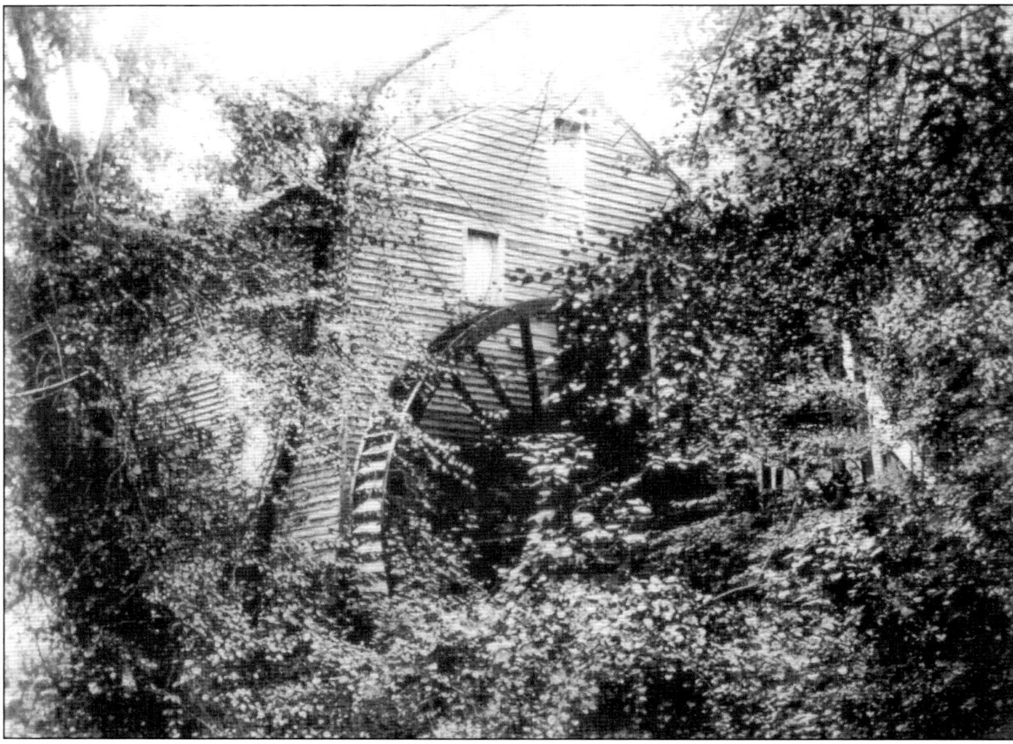

The huge iron waterwheel covered 1.5 stories. When the old mill collapsed following the tornado, the wheel was also damaged. Tannehill Park employees found its axle about half a mile downstream in 1999. The mill pond stayed in place, however, into the 1950s and became a favorite fishing place. The gristmill and dam were rebuilt in 1975 as a tourist attraction, and once again, corn is ground into meal. (Neal Countryman.)

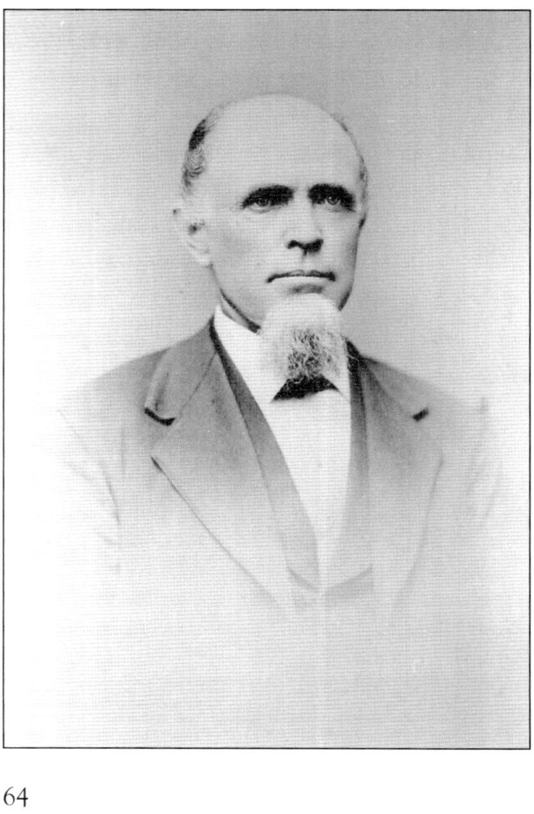

Giles Edwards, a Welshman who operated the Bluff Furnace at Chattanooga, came to Alabama during the Civil War as assistant superintendent of the Shelby Iron Company. After the war, he was employed as a land agent for the Pioneer Mining & Manufacturing Company of Pennsylvania and was living in Tannehill. In 1878, he began construction of a furnace of his own, the Edwards Furnace, in Woodstock. (Virginia Green, Homewood, Alabama.)

This map of the proposed new county of Bessemer, produced by G.W. Chambers for the Bessemer Weekly, dates from 1906 to 1908. The break-away county, which was shot down in the legislature, would have included the Bessemer area of Jefferson County, running through the brown iron ore fields around the abandoned Tannehill Ironworks. It shows many of the old ore mines, including Giles, West Giles, Martaban, Standiford, Greeley, and Goethite. The Edwards Furnace, put into blast by Giles Edwards in 1880, was located near Woodstock, not far from the Giles mine, which he opened for his furnace. (Hueytown Historical Society.)

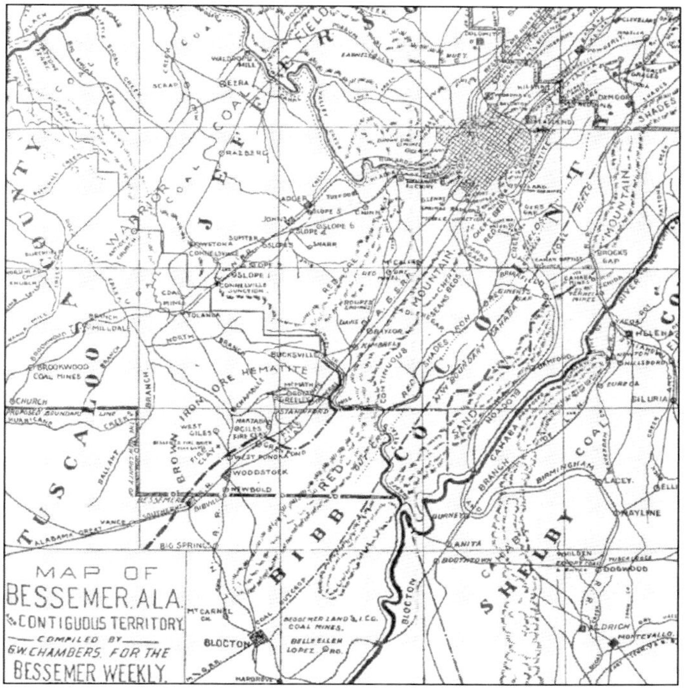

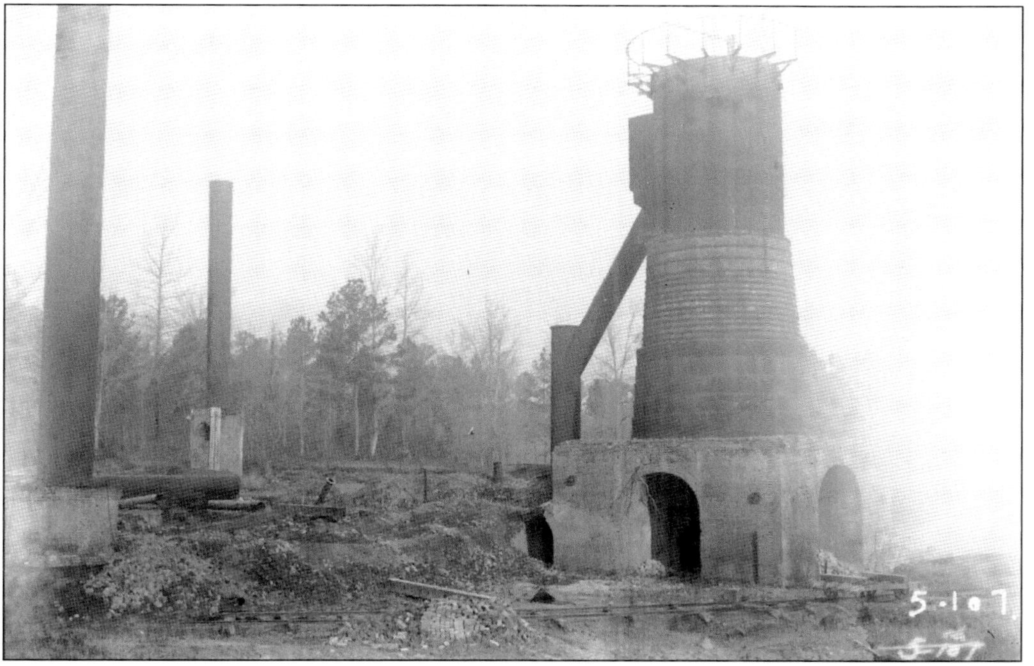

Rebuilt and enlarged several times, the Edwards Furnace was finally blown out in 1890. The plant was dismantled and sold for scrap in 1903, as shown in this picture. The furnace site, still showing some of its few remaining pieces, is today the domain of briars and brambles. Nearby are the tracks of the Norfolk Southern, and the old Giles Ore Mine is underwater. (Eugene A. Smith Collection, University of Alabama Library.)

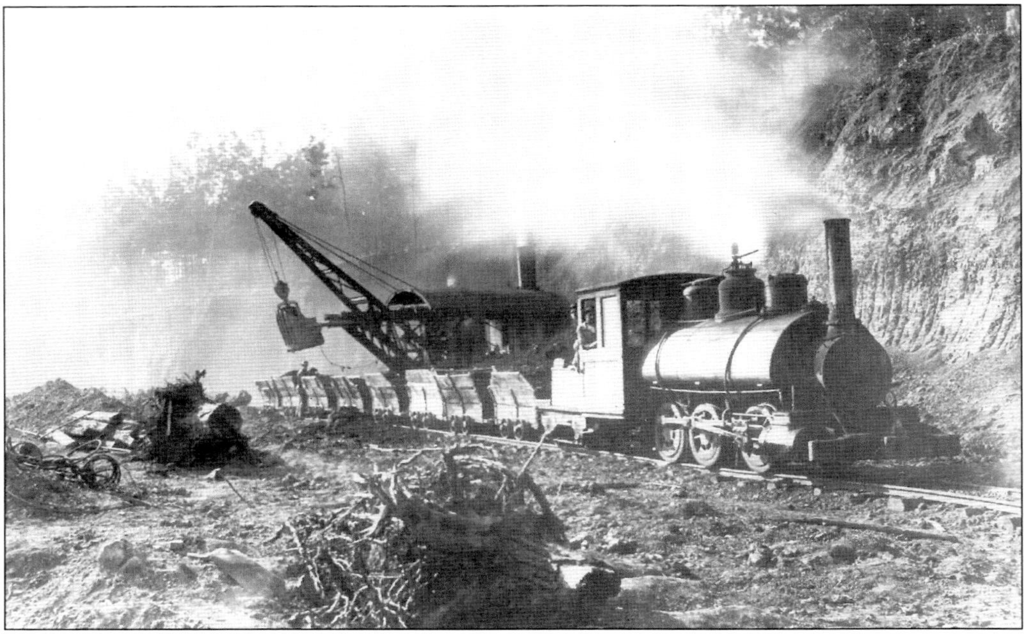

Dummy lines operated by the Pioneer Mining & Manufacturing Company crossed the Goethite Brown Ore Mines between Bucksville and Tannehill in the early 1900s. Steam shovels were used to load the cars along the mineral railroad, and the product was loaded on freight trains headed to Birmingham steel mills at Tannehill Station. (Birmingham Public Library.)

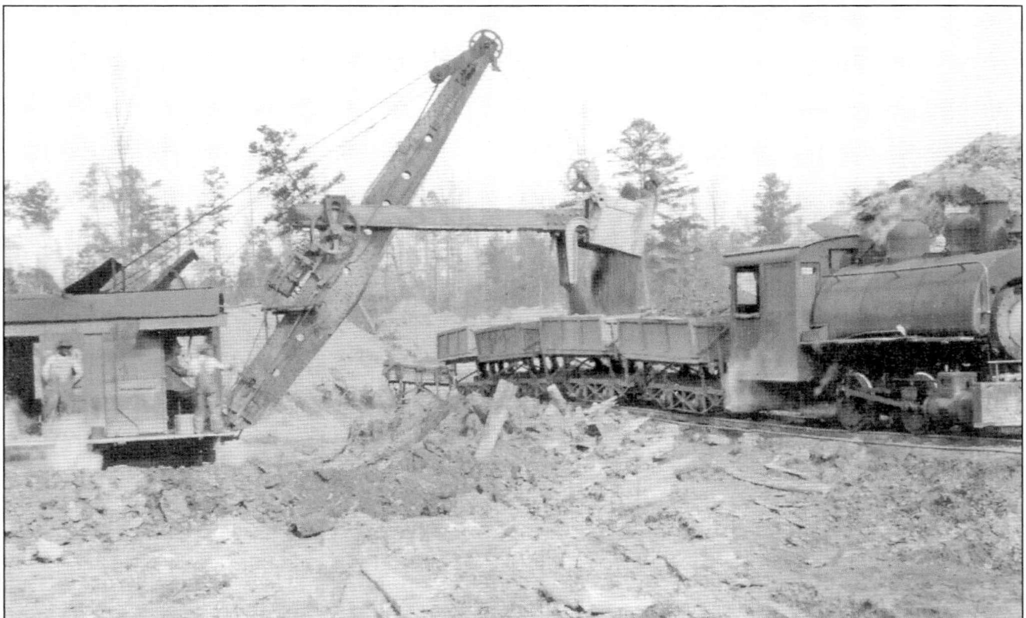

Like hungry monsters, the big steam shovels of mining companies munched away wooded hillsides as they moved tons of brown ore onto the trains. This steam shovel at the Giles Mine of the Central Iron & Coal Company is loading ore on tramcars bound for the washer in 1919. (E.F. Burchard photograph, US Geological Survey Photographic Library.)

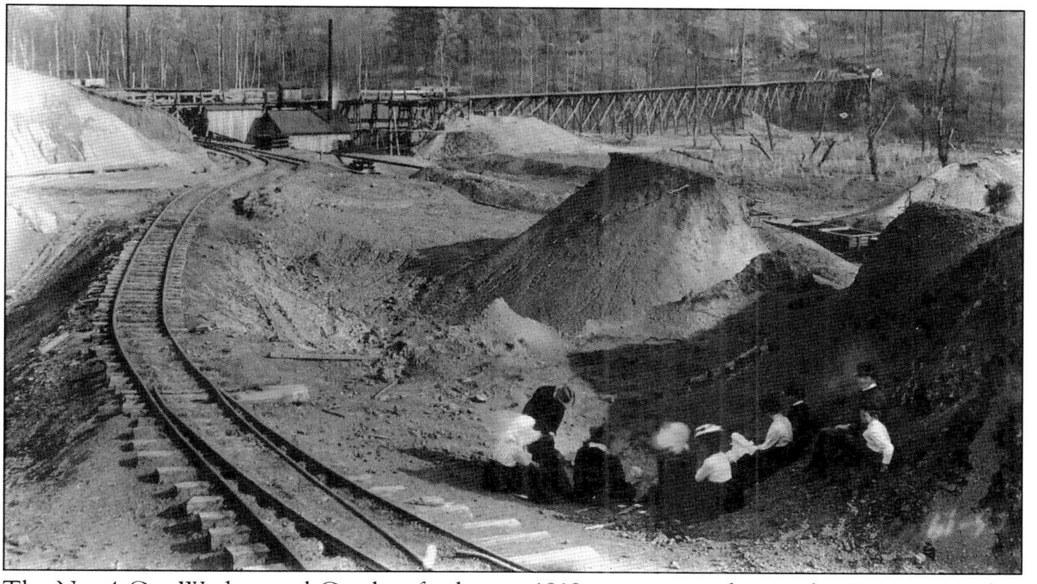

The No. 4 Cut Washer and Crusher facility in 1919 was operated near the Waterton Lakes Subdivision. Although mining here actually began with Daniel Hillman in 1830, postwar mines around Tannehill were opened in 1887 and are now marked by several large lakes where water has filled in mined-out areas. The ore occurs high in the hills, capped with a few feet of red loam and resting against masses of Conasauga limestone. The spent deposits reached 40 to 50 feet in places. Large iron ore boulders, called "dornicks," had to be crushed. Ore was mined here by steam shovel from open cuts that required only stripping. It was also loosened by hand with pick and shoveled and placed into tramcars. The ore was high enough to be trammed by gravity from many of the cuts to the crusher and washer. (Tutwiler Collection, Birmingham Public Library.)

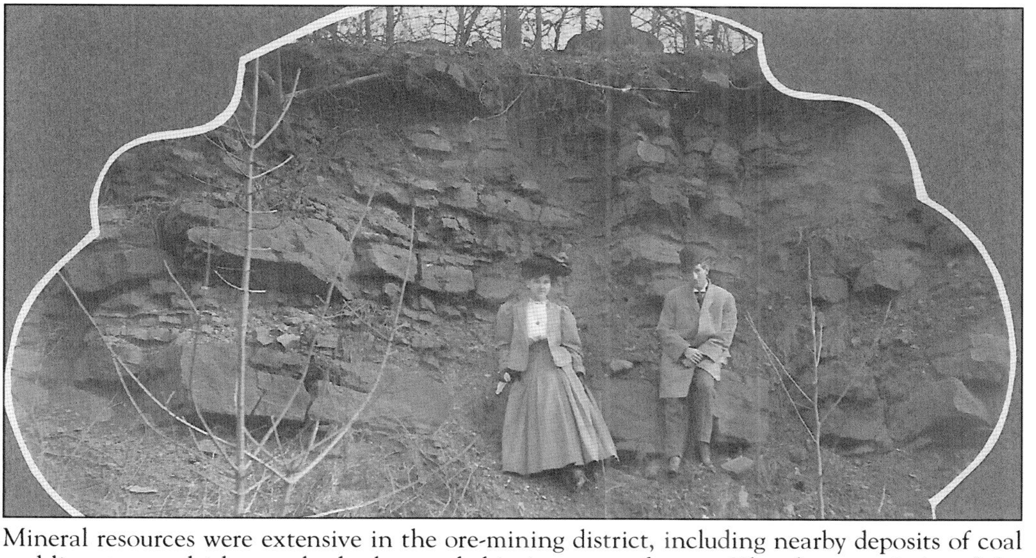

Mineral resources were extensive in the ore-mining district, including nearby deposits of coal and limestone, which were both also needed in iron manufacture. The close proximity of the minerals cut down on transportation costs and meant iron made in the Birmingham district could be made cheaper than almost anyplace on the globe. This c. 1910 coal deposit shows a carefree couple taking in the sights.

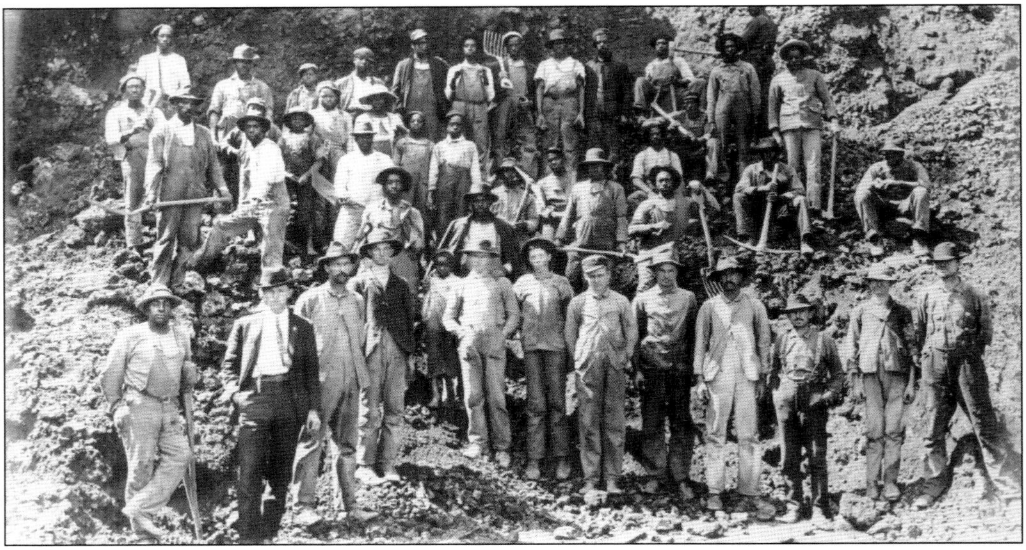

The Pioneer Mining & Manufacturing Company, the first of the big ore companies in the Tannehill area after the Civil War, hired hundreds of day laborers to work in its mining operations, including these ore miners at the Houston Mine (above) and stone cutters at Dale (below). These photographs are dated around 1900. The workforce, which cut across racial lines, arrived daily on the trains and drew workers from Jefferson, Bibb, and Tuscaloosa Counties to various company sites. From the 1880s through the 1920s, the demand for mineral resources, including iron ore, coal, and limestone, remained strong because of the increased production in iron, cast iron pipe, and manufactured products. (Both, Republic Iron & Steel Company Photographic Album, Archives Department, Birmingham Public Library)

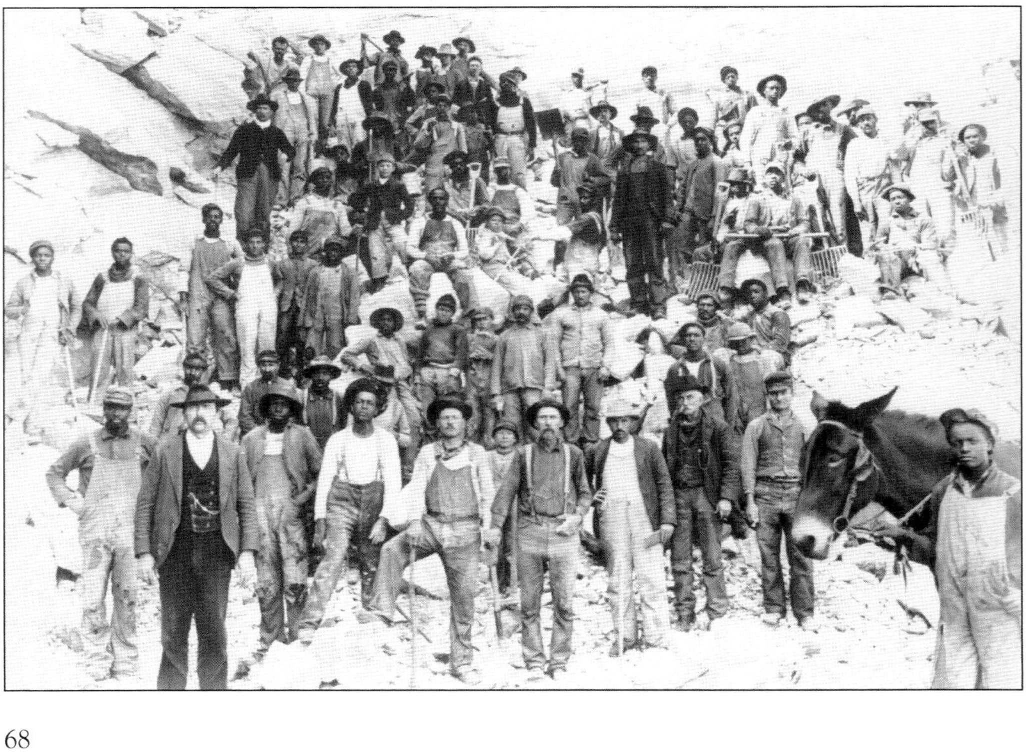

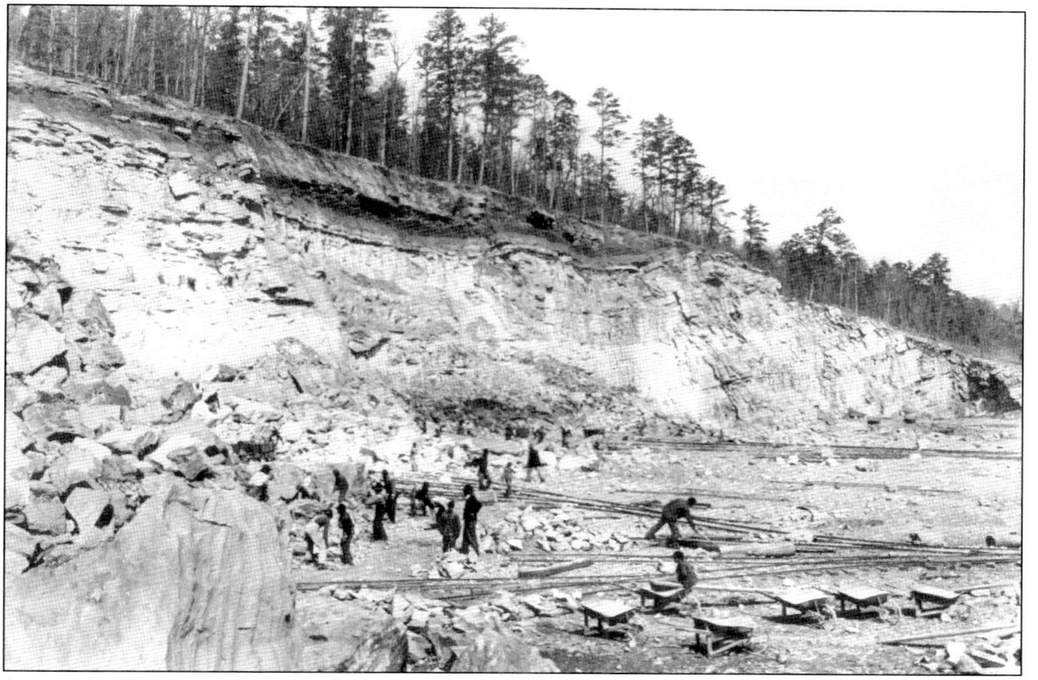

Limestone deposits were identified around Tannehill and in neighboring Jefferson County. This mine at Dale (Village Springs) was owned by the Pioneer Company. Here, workers break loose chunks of limestone that will be fed into a crusher. Limestone, as a fluxing agent, is used in the iron-making process to remove dirt, silica, and foreign matter as slag. The rail carried steam shovels and open cars to limestone mining areas. (Tutwiler Collection, Birmingham Public Library.)

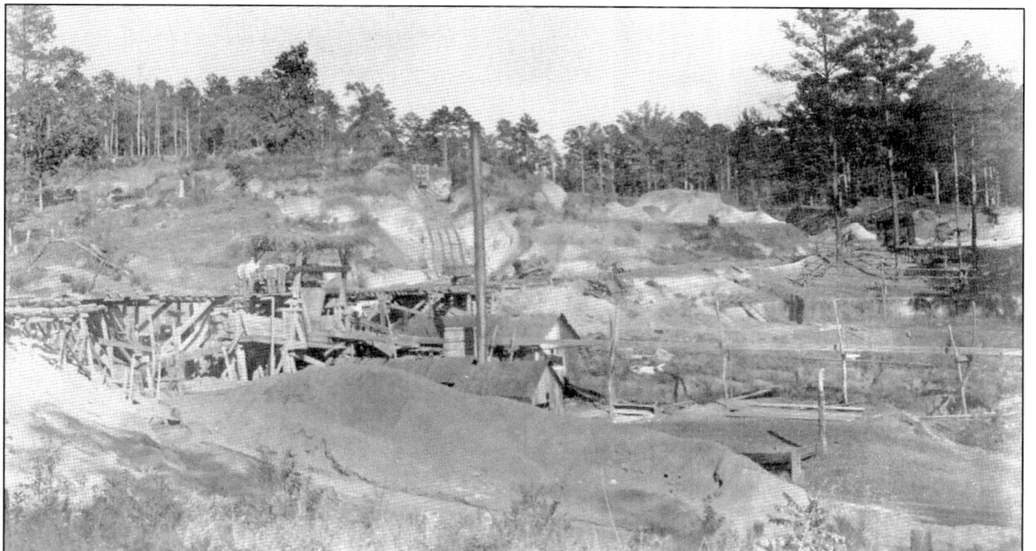

The concrete footprint and some rock foundations for the washer plant at the No. 4 Cut, operated by the Pioneer Mining & Manufacturing Company, are still visible, including the water channel. Thousands of tons of brown iron ore were cleaned here daily before the demand for Alabama ore lessened with higher-grade ore imports supplying local steel plants. (Birmingham Public Library.)

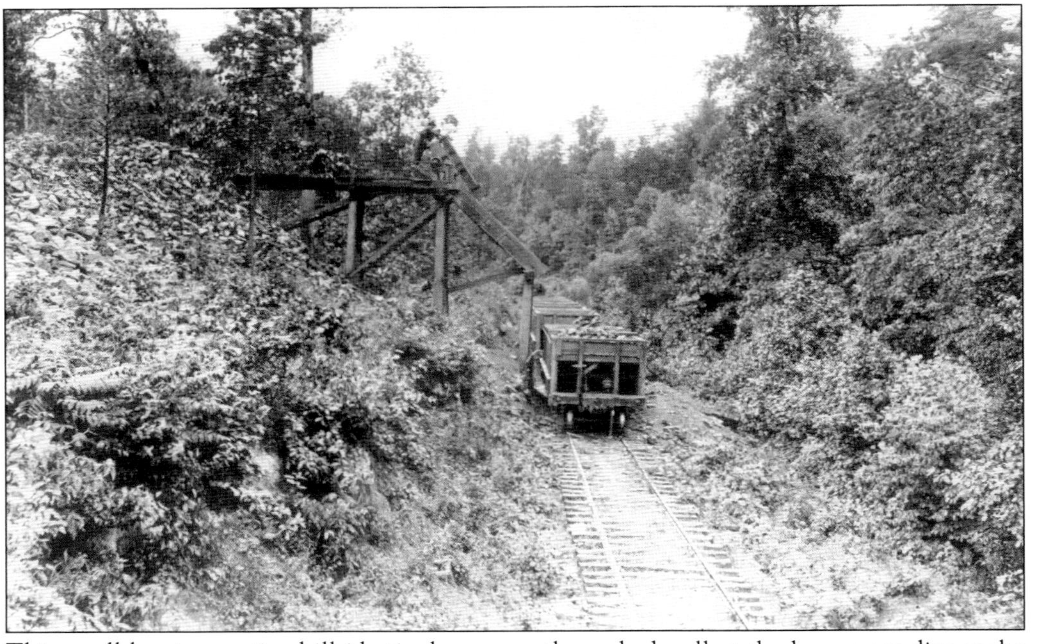

This small but innovative hillside tipple apparently worked well at the brown ore digs at the Debardeleben Mine, operated by the Pioneer Company. Note the dump car and shoot discharging a load into cars along the Mineral Railroad. Some of the ore mines were in the far reaches of mining property. (Birmingham Public Library.)

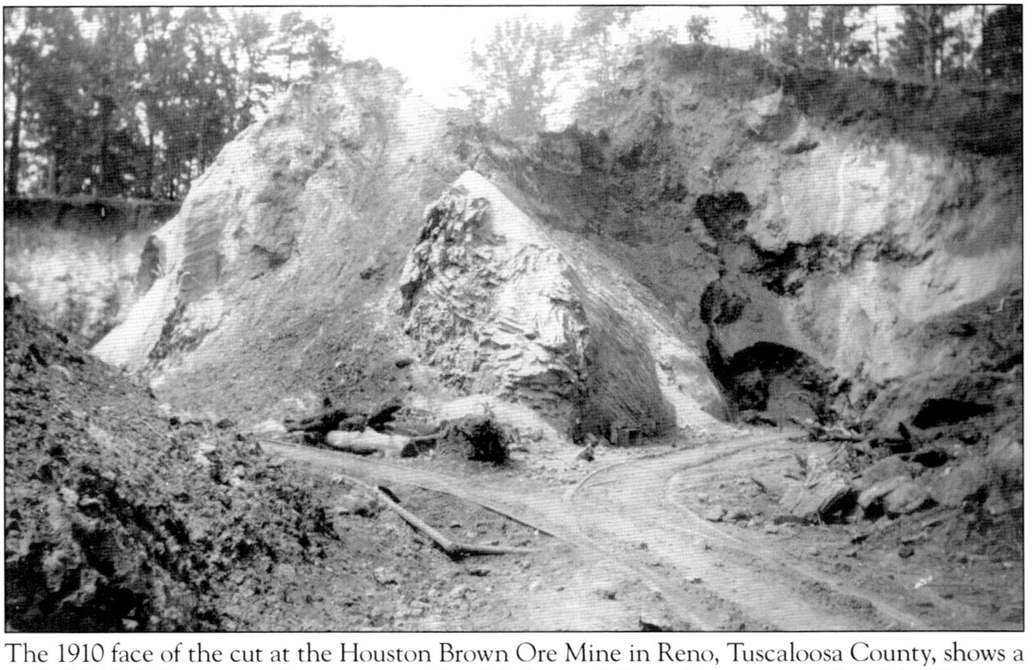

The 1910 face of the cut at the Houston Brown Ore Mine in Reno, Tuscaloosa County, shows a reef of folded Conasauga limestone, residential clay on left, and iron ore on the right. Ore was extracted from this site to feed Birmingham and Tuscaloosa blast furnaces in the late 1890s and early 1900s. (US Geological Survey Photographic Library.)

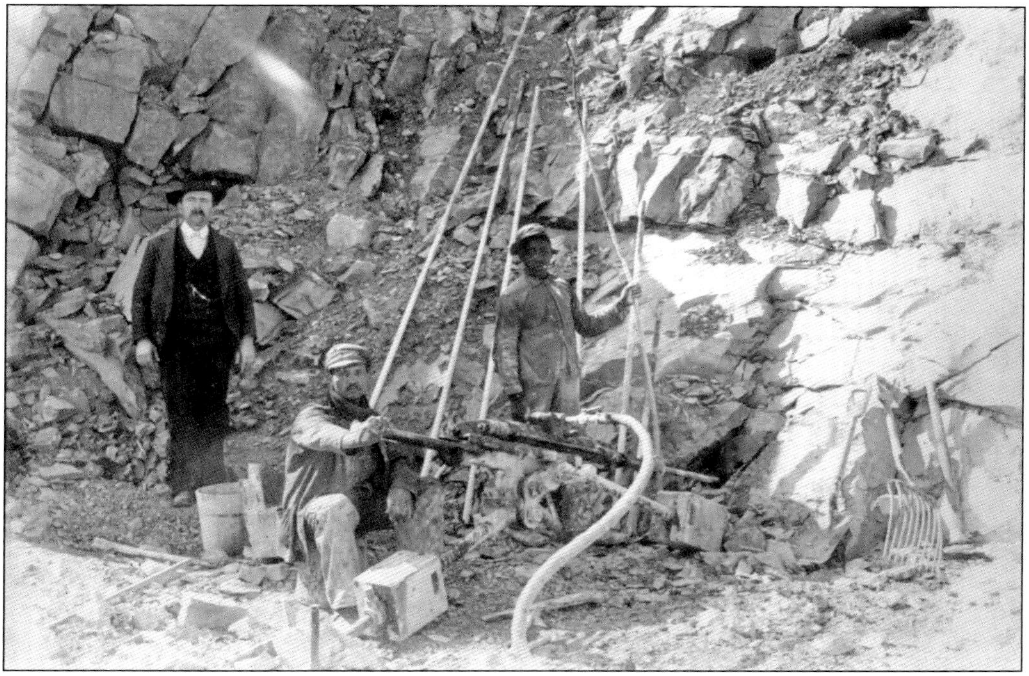

A miner handles a steam-powered drill on a limestone bank at the Pioneer Company's mine at Dale in Jefferson County. Limestone was used as a flux to remove impurities from iron ore in the reduction process at blast furnaces. Tamping rods were used to pack explosives into the drill holes. (Birmingham Public Library.)

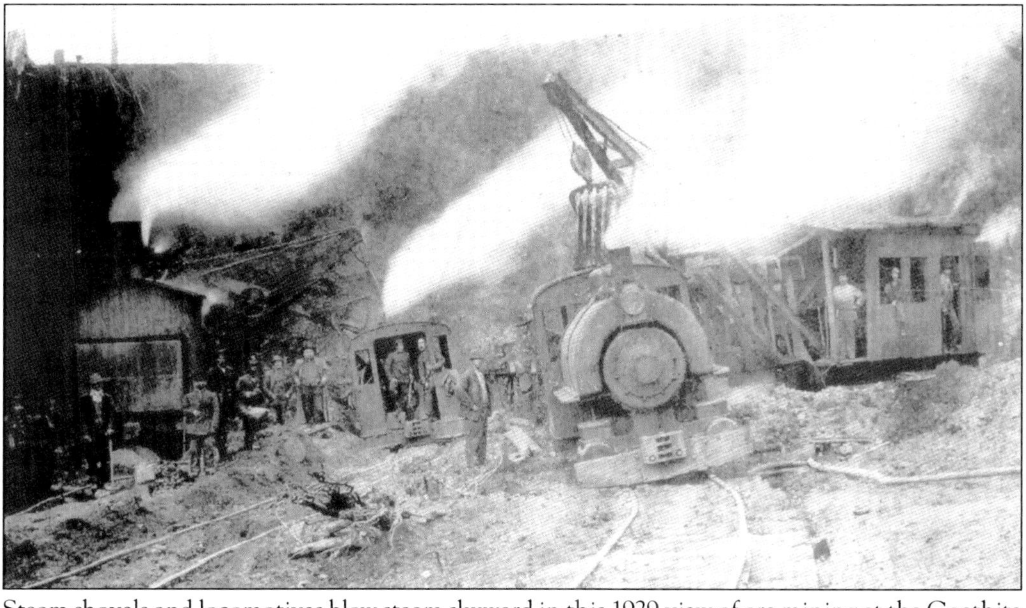

Steam shovels and locomotives blow steam skyward in this 1929 view of ore mining at the Goethite deposit near Tannehill. This image is evidence of the heavy ways of toil in extracting iron ore. A commissary was located near the mine for use by workers, and trains ran on schedule from the Tannehill Station bringing workers to and from work.

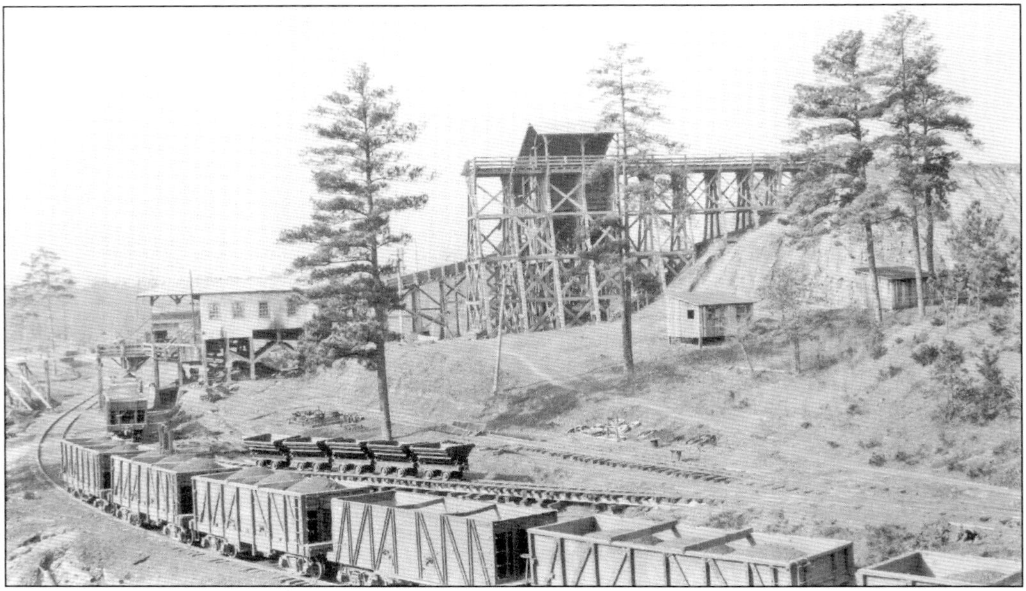

On December 13, 1929, an ore train prepares to leave the brown ore washer of the Central Iron & Coal Company at Giles in Bibb County. The high tipple in the background speaks to the size of the operation. Much of the ore mined here was shipped to the Holt Furnace, located north of Tuscaloosa. (E.F. Burchard photograph, US Geological Survey Photographic Library.)

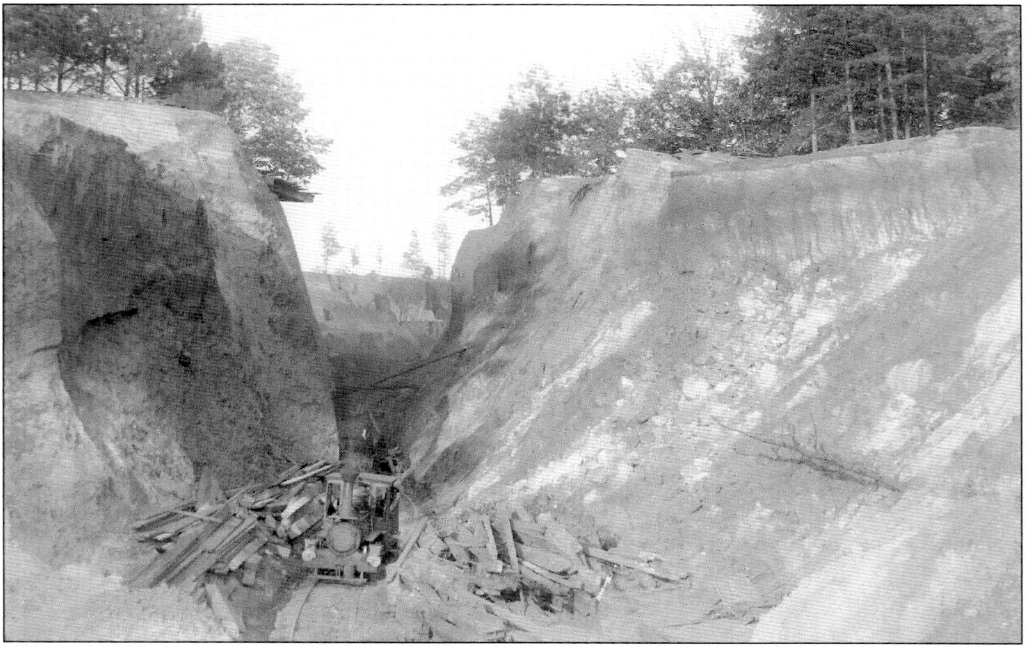

Brown iron ore reserves were extensively mined in Tuscaloosa, Bibb, Jefferson, and Shelby Counties. This photograph from the 1880s shows a deep cut in a Shelby County mine, which provided ore for the Shelby Iron Company after the Civil War. The ore vein has been mined all the way down to the railroad grade. It appears the photograph was taken after a bridge collapsed at the top of the hill. Note the ancient engine utilized in the clean-up.

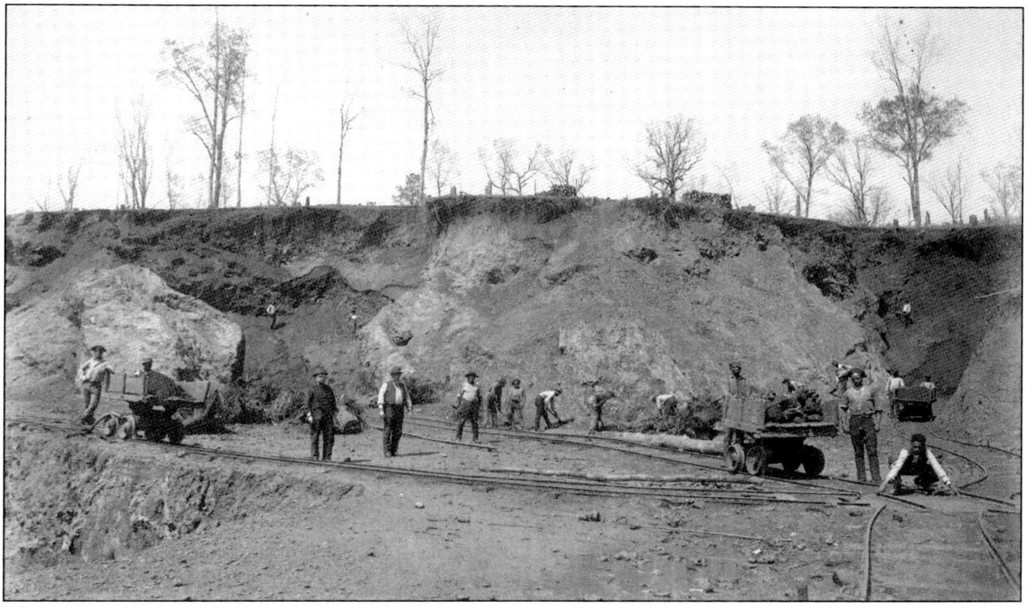

The mines of Tannehill, among the first opened around 1887–1893, were located on the north side of the railroad toward the present-day site of the Tannehill Baptist Church. In this photograph, brown ore miners lay track ahead of the arrival of a steam shovel, which will push the hillside further back. (Alabama Department of Archives & History.)

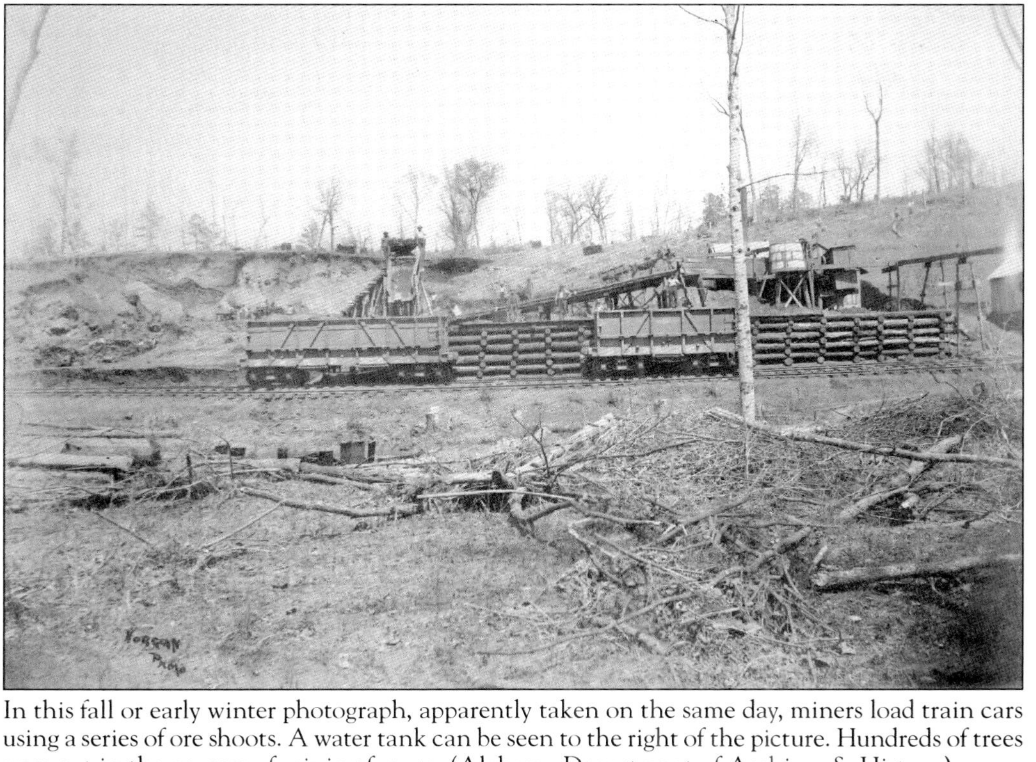

In this fall or early winter photograph, apparently taken on the same day, miners load train cars using a series of ore shoots. A water tank can be seen to the right of the picture. Hundreds of trees were cut in the process of mining for ore. (Alabama Department of Archives & History.)

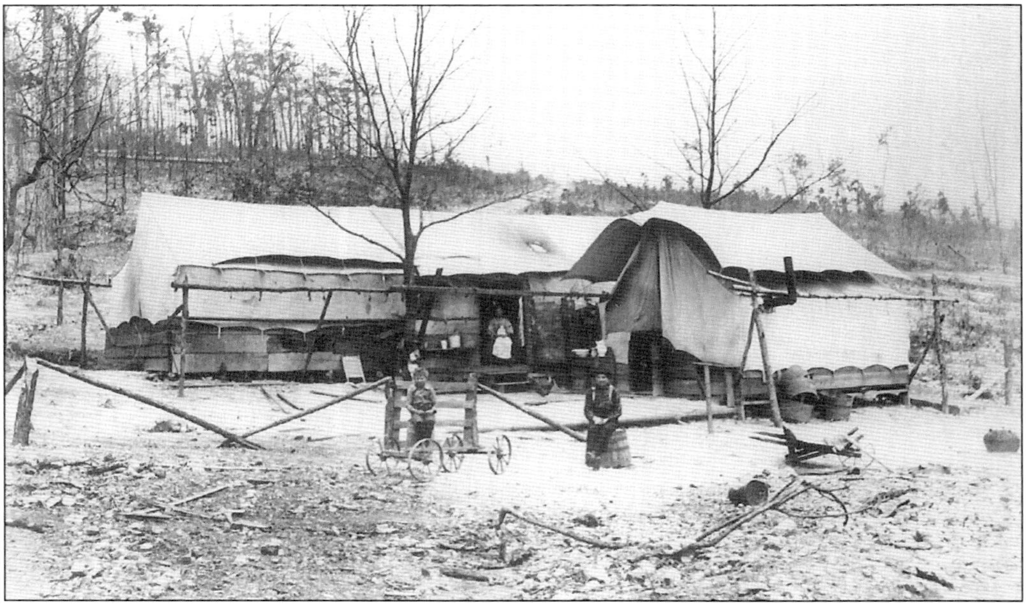

Some early miners, shown around 1905, lived in tents until wood-frame or log houses could be built, like this one near Tannehill. While many miners traveled by train to work, others lived in shacks in the area. It is not known what worker housing was provided at Goethite; however, concrete privy foundations could be seen on the Russell Farm into the 1980s, perhaps as part of its commissary or a bathhouse. (Southern Collection, Birmingham Public Library.)

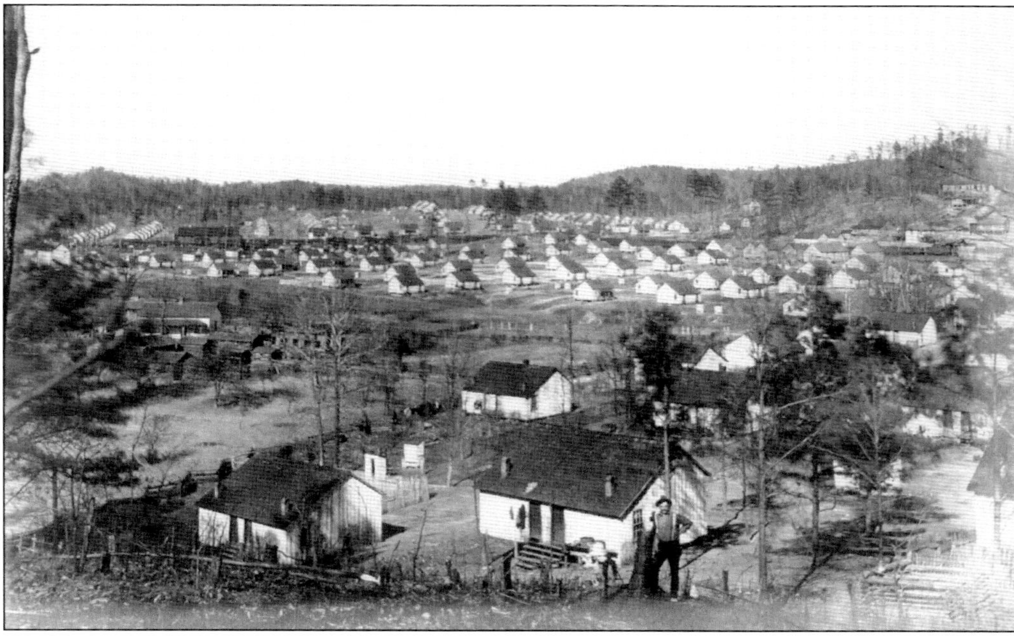

Worker housing extends to the horizon in this mining camp photograph taken of the Pioneer Mining & Manufacturing Company's property at Republic. Brown ore mines at Tannehill, Goethite, and Houston and the coal mines at Sayreton and Republic were opened under John Adam's supervision for the Pioneer Company from 1892 to 1899. (Birmingham Public Library.)

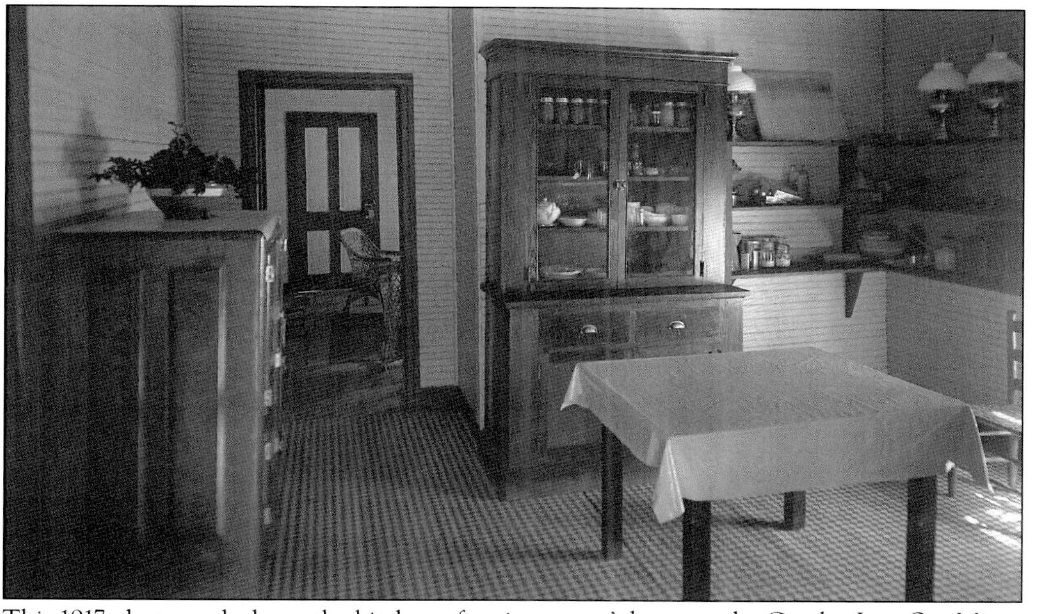

This 1917 photograph shows the kitchen of an instructor's home at the Greeley Iron Ore Mines west of Tannehill, near present-day Caffee Junction. The Tennessee Coal, Iron & Railroad Company (TCI) not only built housing for workers but also provided schools and medical facilities for both blacks and whites. The houses set aside for teachers were neat and well furnished. (Land Department, TCI.)

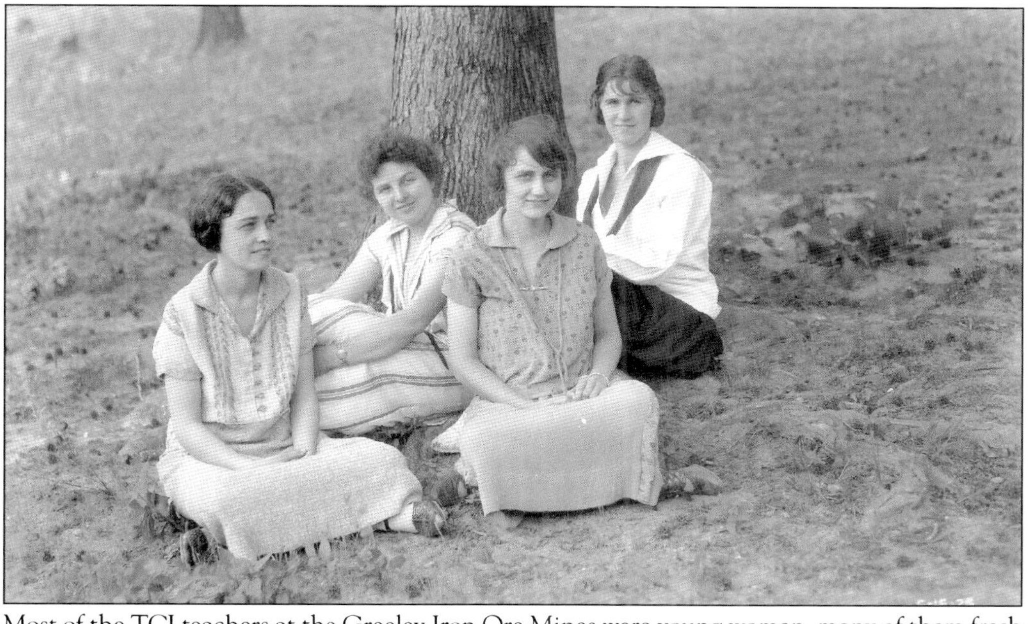

Most of the TCI teachers at the Greeley Iron Ore Mines were young women, many of them fresh out of college. These teachers of the white school pose by a tree near the Greeley School in 1925. By 1900, TCI was operating 15 iron ore mines in the Birmingham district. In 1914, TCI began to provide facilities and programs to improve lives of employees with schools, hospitals, training programs, and housing. (Land Department, TCI.)

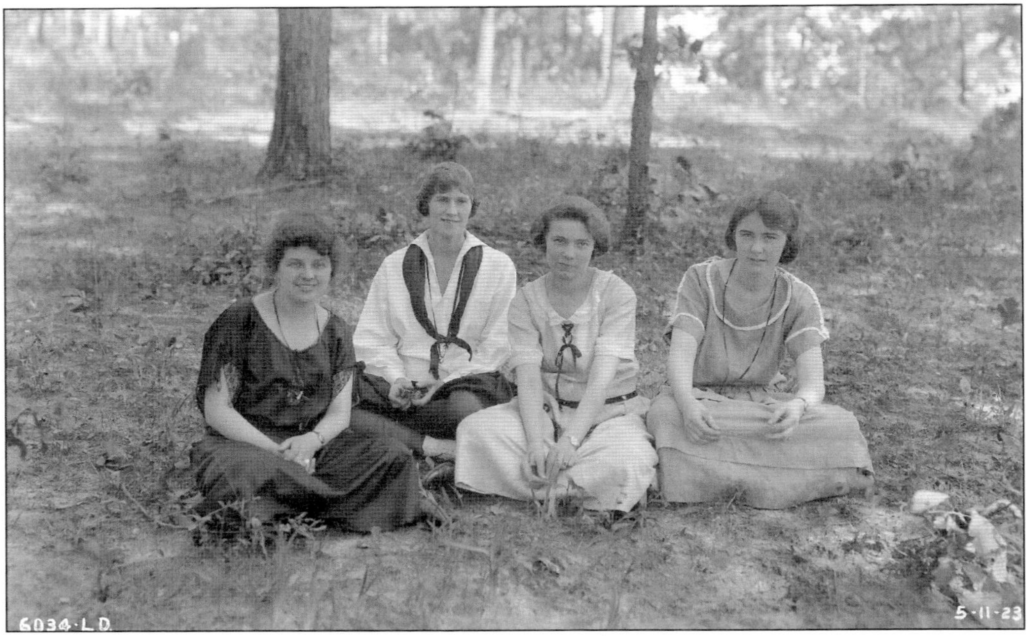

This group of Greeley Iron Ore Mine teachers was on the job in 1923. Many young instructors favored working for steelmakers rather than in the public school system due to the benefits of housing and medical care. Those pictured are, from left to right, Emily Roberts, Nell Burdick, Louise Hinton, and Virginia Hart. (Land Department, TCI.)

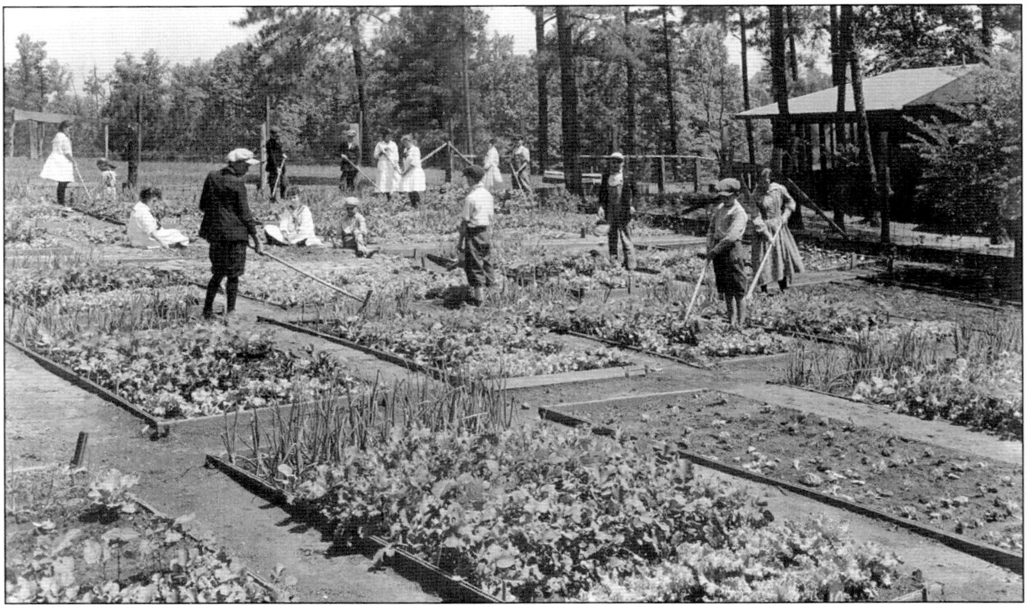

Gardening instruction was stressed, and each school had a demonstration garden behind it. The growing area was divided into plots with prizes awarded to the best crop based on preparation, fertilization, cultivation, and appearance. Students raised mustard, lettuce, cabbage, beans, onion, peas, radishes, and beets. This photograph shows the Greeley White School garden in 1920. (Land Department, TCI.)

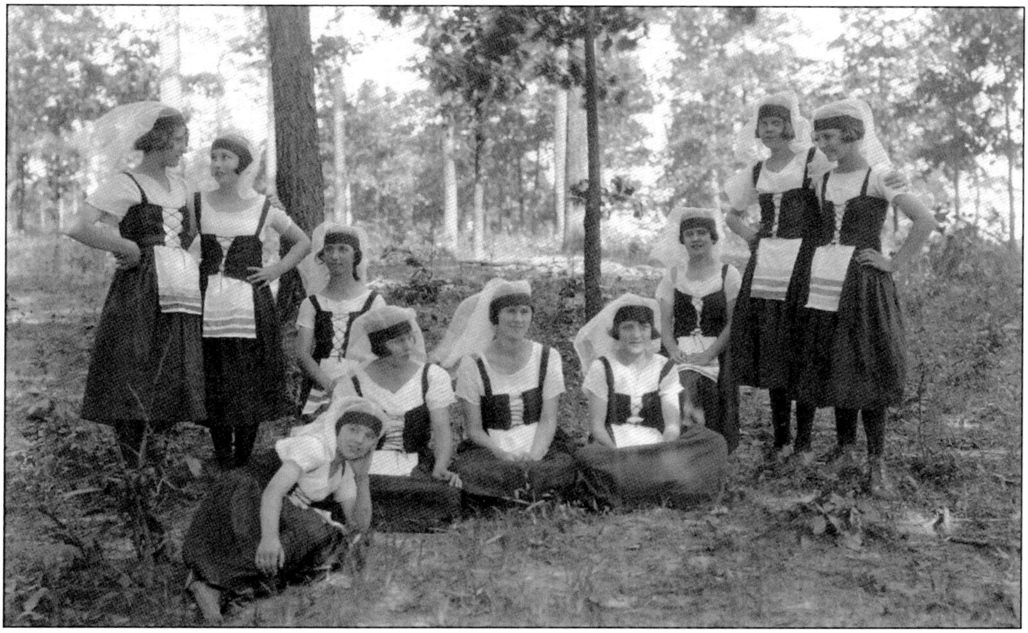

The theatrical arts were also stressed, with students participating in various plays and pageants. Here, students from the Greeley Mine School participate in the Swedish dance as a part of the white school's pageant, called *The Conflict*, which was held at Muscoda in 1922. Muscoda was a red ore mining camp. (Land Department, TCI.)

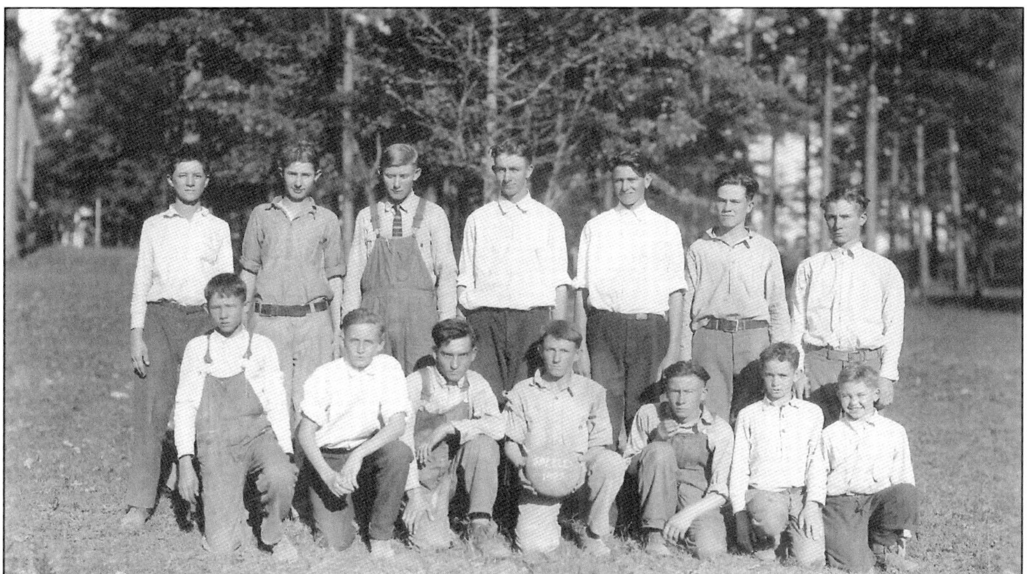

The Greeley Iron Ore Mines boys' speedball team poses for this picture in 1925. Although 11 players constitute a regulation team, the game can be played with fewer members on each team. A wide variety of techniques are employed in speedball, including kicking and dribbling the ball with the feet as in soccer, catching and throwing the ball as in basketball, and punting and passing the ball as in football. (Land Department, TCI.)

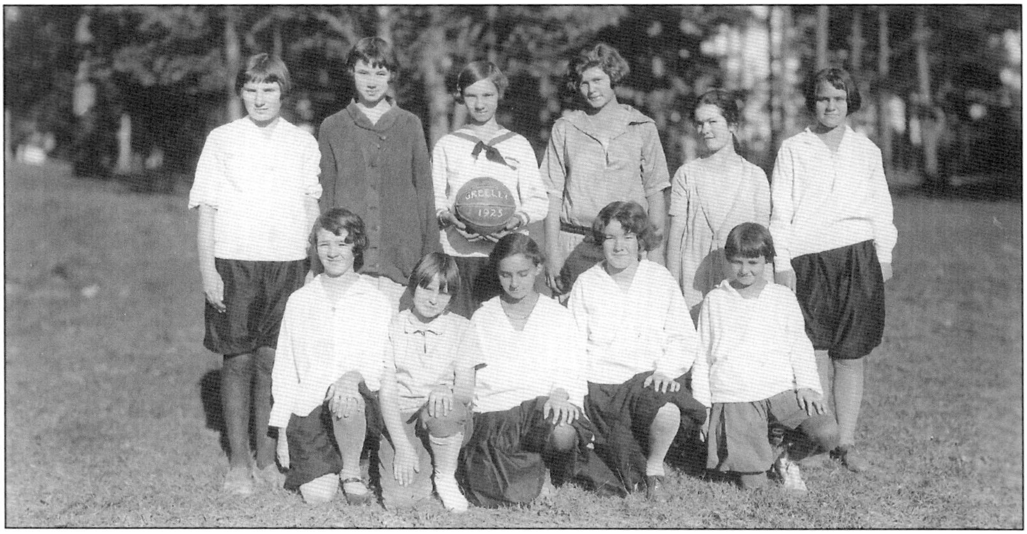

The Greeley girls' volleyball team is shown here, also in 1925. Boys and girls were encouraged to participate in sporting events, as well as gardening and theater. Mining camp schools often competed against each other in sports and other school activities. Many students lived in company housing. (Land Department, TCI.)

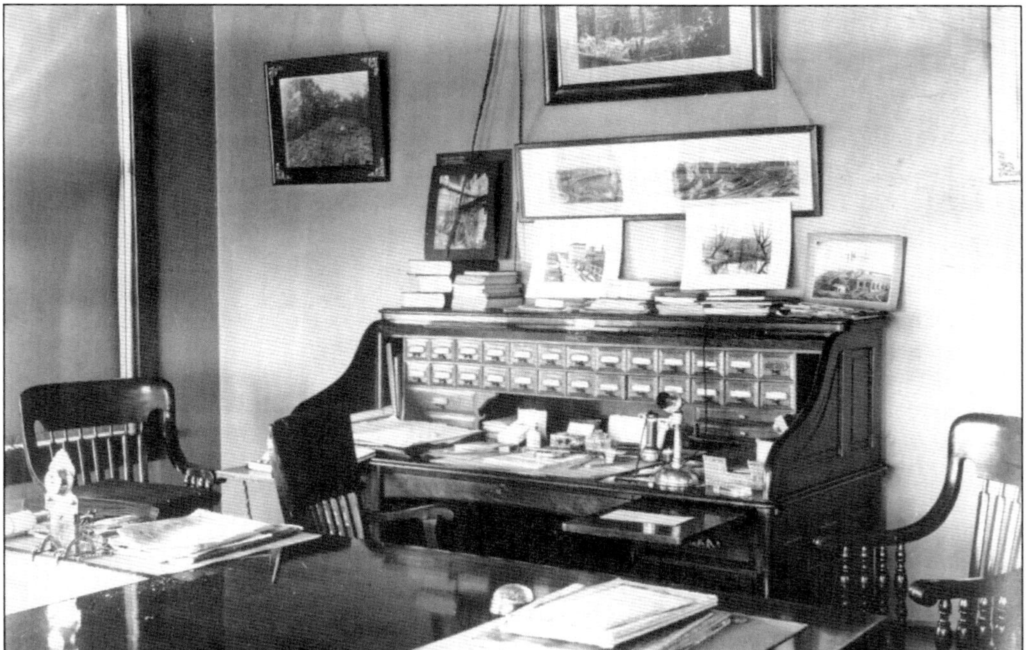

This is a picture of the manager's office at the Republic Iron & Steel Company's Thomas Works in Birmingham about 1920. It shows a number of photographs on the wall, including a large picture of the old Tannehill iron plant then owned by Republic. Although the Tannehill mill lay dormant, Republic continued to mine large amount of brown iron ore nearby. The Tannehill picture on the wall is now in the Special Collection Library at Samford University. (Birmingham Public Library.)

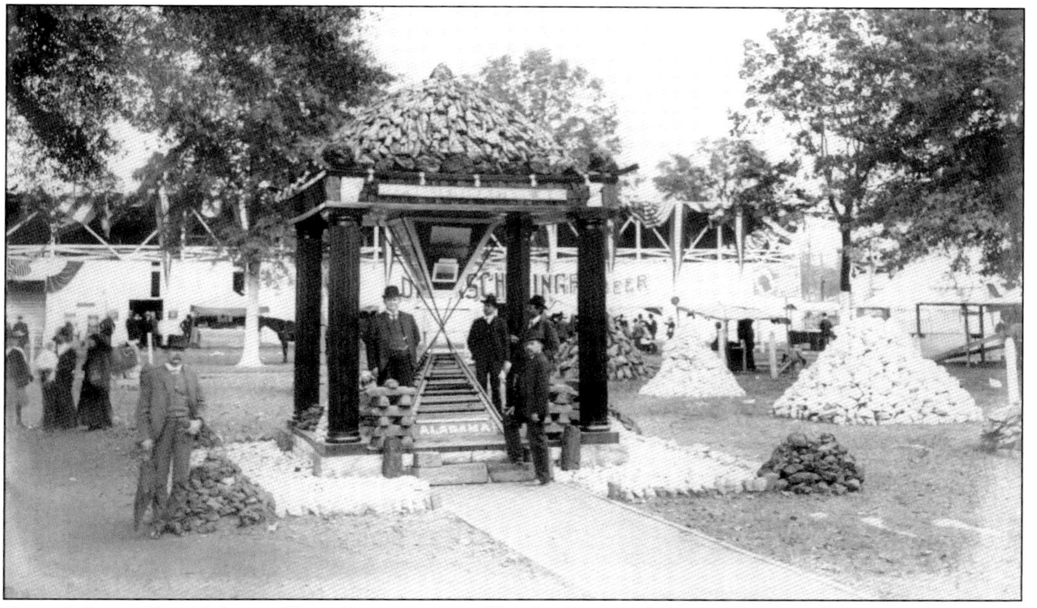

An exhibit at the Alabama State Fair, sponsored by the Pioneer Mining & Manufacturing Company, the Birmingham Rail Manufacturing Company, and Republic Iron & Steel Company, features coal, pig iron ingots, iron ore, and limestone from Republic's various properties. The Vulcan statue appeared at the 1906 Alabama State Fair, about the time of this picture. It previously had been at the 1904 World's Fair in St. Louis. (Republic Iron & Steel Company Collection, Birmingham Public Library.)

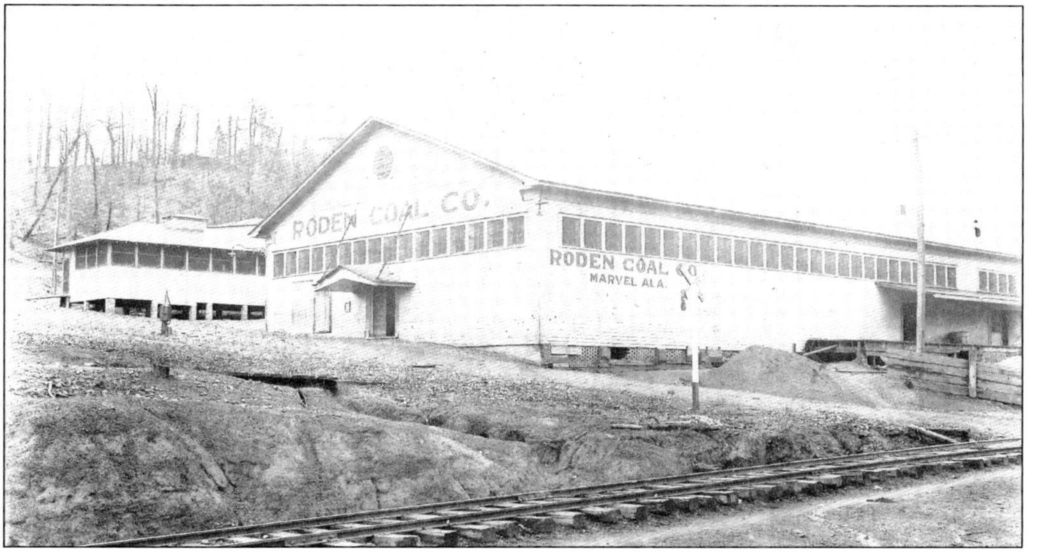

The Roden Coal Company Commissary, shown around 1910, was located at Marvel, 25 miles south of the Tannehill Furnaces. This facility was one of the many properties of Benjamin F. Roden, a pioneer Birmingham investor. He built Birmingham's first streetcar railway and founded the town of Avondale and the Hillman Hotel Company. Coal from his Marvel Mine was marketed far and wide. Commissaries of this type were common in large mining camps. (R.H. Horton photograph.)

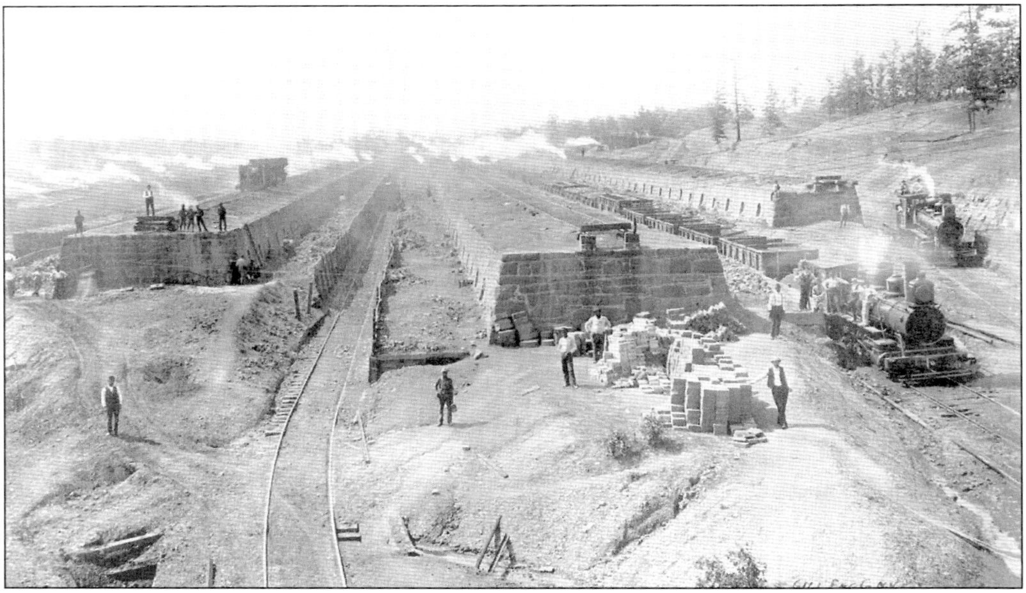

This is a view of the Blocton Coke Ovens from the north end. The men in the foreground are standing next to a stack of large refractory brick needed to make frequent repairs. Some 467 beehive coke ovens were built here, many of them still standing. (*Tennessee Coal, Iron & Railroad Company Annual Report*, 1900.)

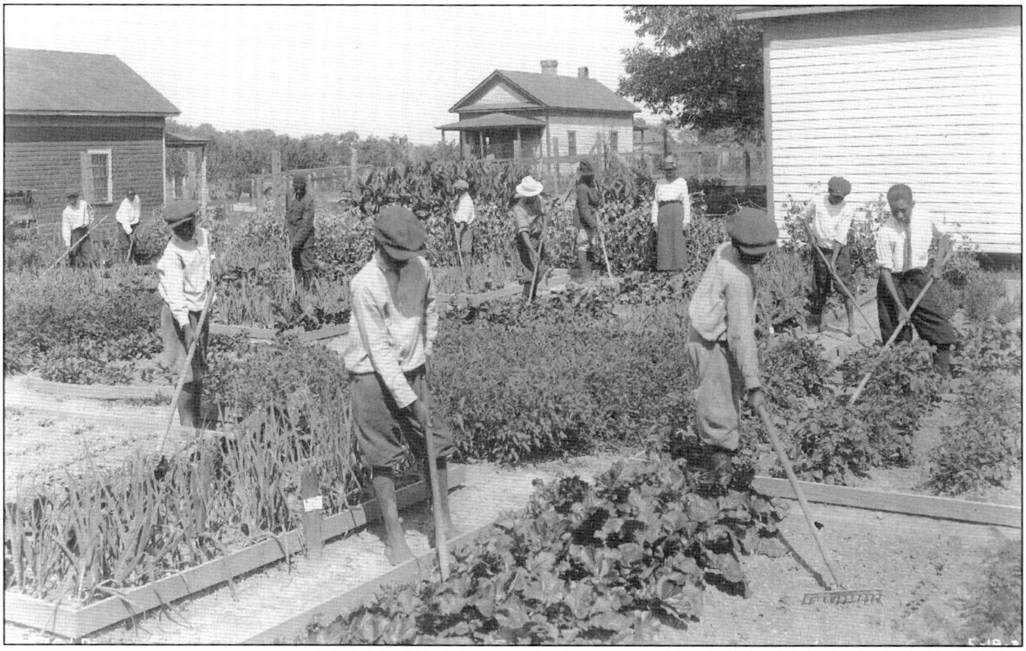

This group of students at the Ishkooda Ore Mines Black School won first prize for their effort in 1921. Ishkooda was a red ore mine on Red Mountain, near Birmingham. The Tennessee Coal, Iron & Railroad Company also operated it. Note the company housing in the area, which is now near Red Mountain Park. (Land Department, TCI.)

Five

Reclamation and Recovery

The historic iron furnaces in Roupes Valley, a landmark since the Civil War, quickly became a favorite haunt for the curious. Beginning with a major archaeological study by the University of Alabama in 1956, they have been studied more closely than any other ironworks in the South.

Since then, six more archaeological teams have made their own site excavations, each issuing reports that turn back the pages of the past. Hundreds of artifacts have been recovered, from pig iron ingots and cooking ovens to cannonballs and machine parts.

Almost immediately after the furnace yard was turned over to Tannehill State Park in 1970, work got underway preserving the remains of Furnace No. 1, which had been dynamited in 1968, presumably by people who did not want their hunting grounds turned into a tourist attraction.

University of Alabama officials first suggested the site be protected and restored in 1952. While stabilization work began in 1970, it was the American Bicentennial in 1976 that spurred an investment of $127,000 in repairs to make the 1859 furnace capable of being put back into production. The work was done under the supervision of engineer Max Hale. To continue the restoration, the park allocated another $112,000 to rebuild the fallen front sections of Furnaces Nos. 2 and 3, erected in 1862. This work took place between 1985 and 1987. The heavy masonry work was conducted by Edwin Robinson from Cullman, Alabama, under the direction of architect Ken Hester. In 2005, architect Vic Fortinberry directed reconstruction of the bridge houses and charging bridge. The work continues.

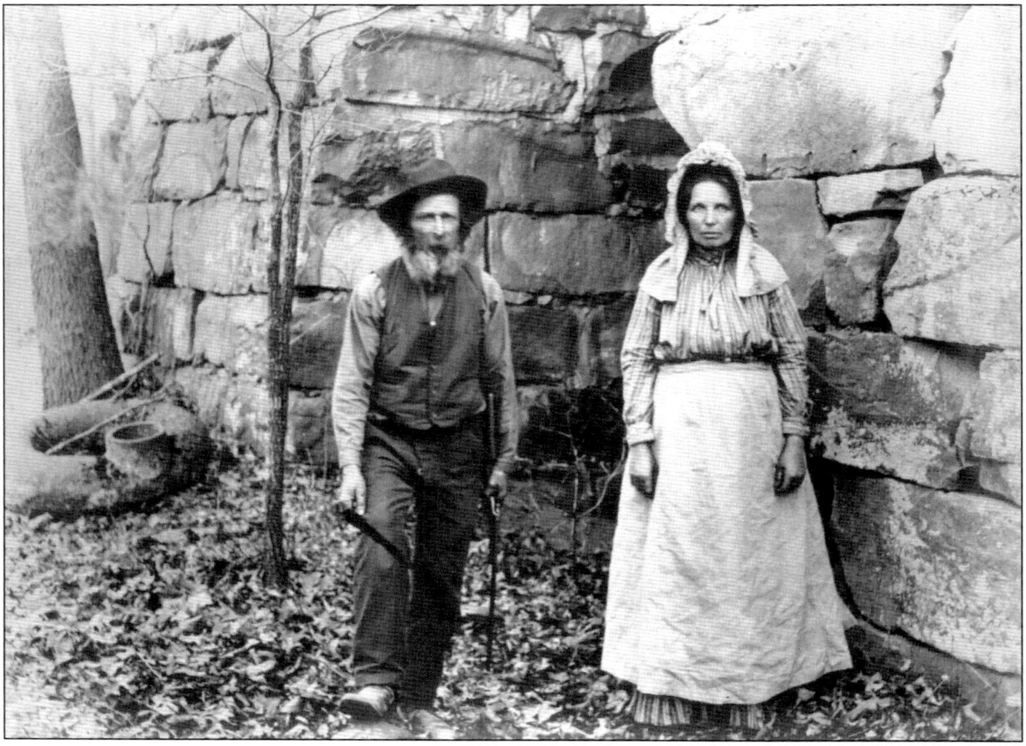

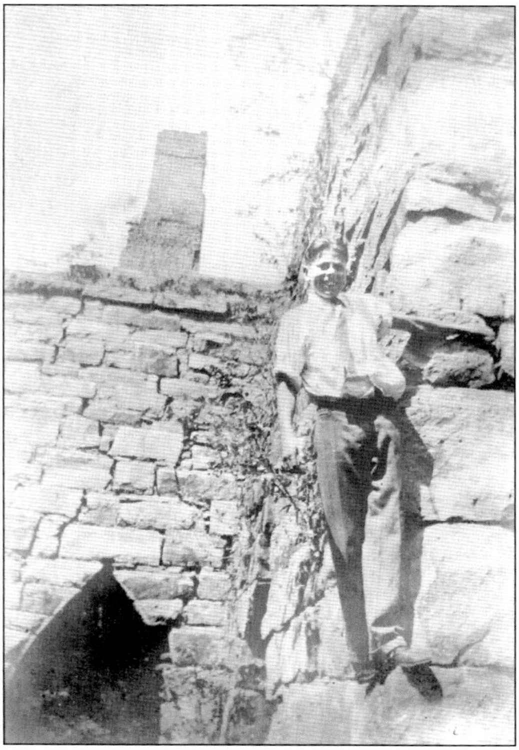

The Tannehill Furnace ruins became a favorite spot for hunters around the turn of the century. William M. Hamaker (1841–1912) and his wife, Mary Jane (1859–1939), are shown here in the late 1890s. Hamaker, whose homestead was about two miles southwest of the works, was once employed there. A furnace fire ring is in the left-hand corner of the photograph. (Republic Iron & Steel Company Collection, 1888–1910, Birmingham Public Library.)

From his high perch, 11-year-old C.A. Smithson of Bessemer got a good view of things at Tannehill Furnace No. 1 in 1924. Atop Furnace No. 2, the old brick draft stack was still in place in this photograph. A bridge house originally covered all but the top of the chimney. (C.A. Smithson, Bessemer, Alabama.)

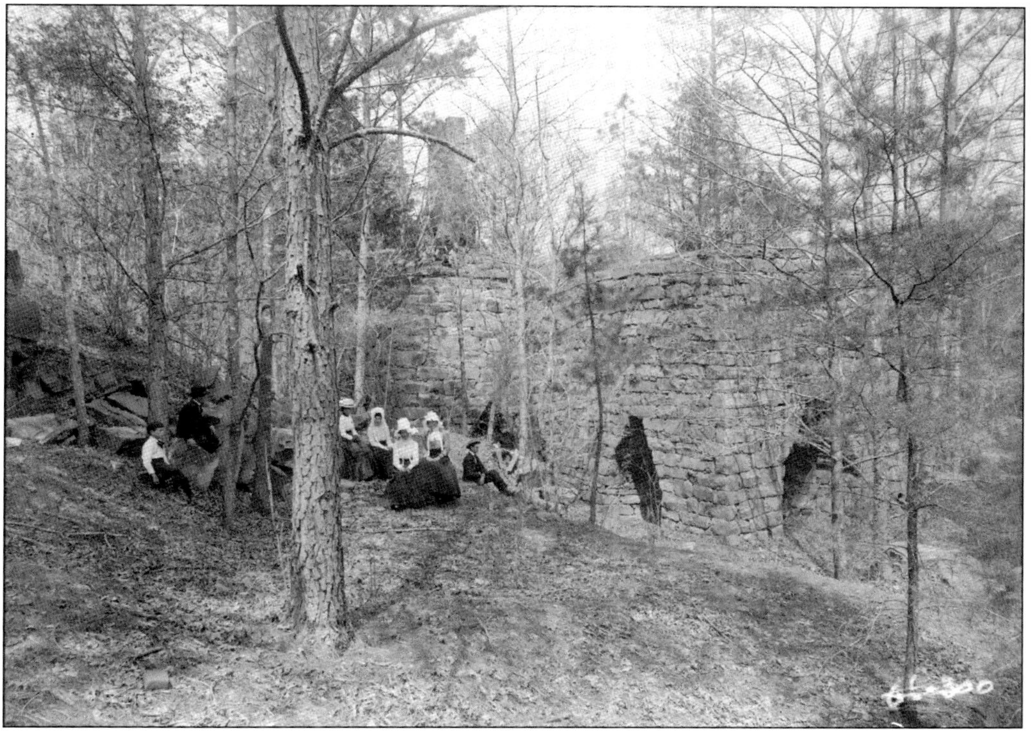

On this day trip in the 1890s, men, women, and children in their Sunday best take in a bit of history behind the three Tannehill Furnaces. A more adventuresome group sits atop the double furnaces beneath a brick draft stack that once blew smoke from fires that ranged between 2,600 and 3,100 degrees. (Alabama Department of Archives & History.)

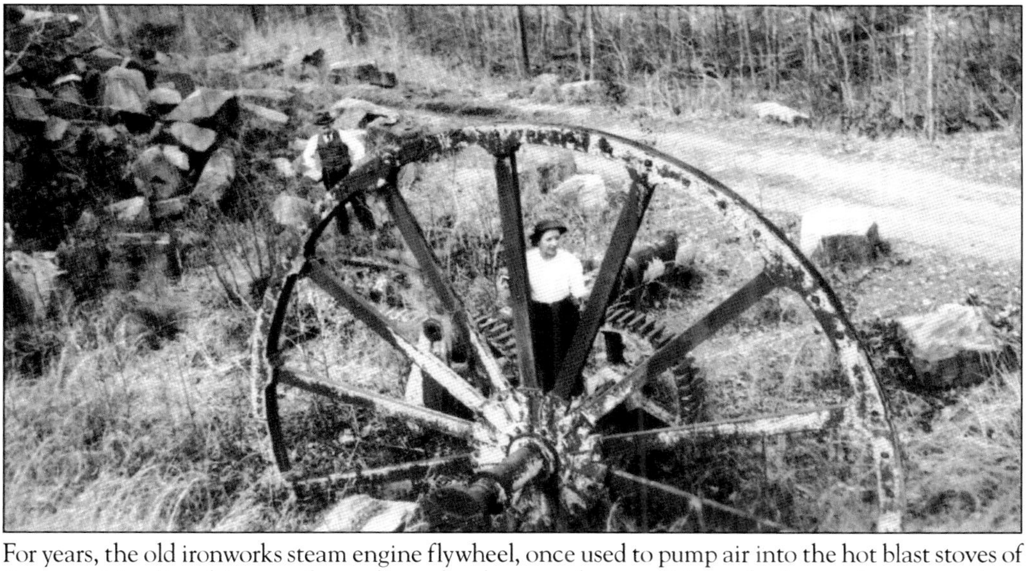

For years, the old ironworks steam engine flywheel, once used to pump air into the hot blast stoves of Furnaces Nos. 2 and 3, was a favored spot for photographers until it was broken up by scrap seekers during World War II. It is still in place in this 1941 picture, when this couple came for a winter visit. The wheel was 15 feet in diameter, the bottom third of which is still in the ground.

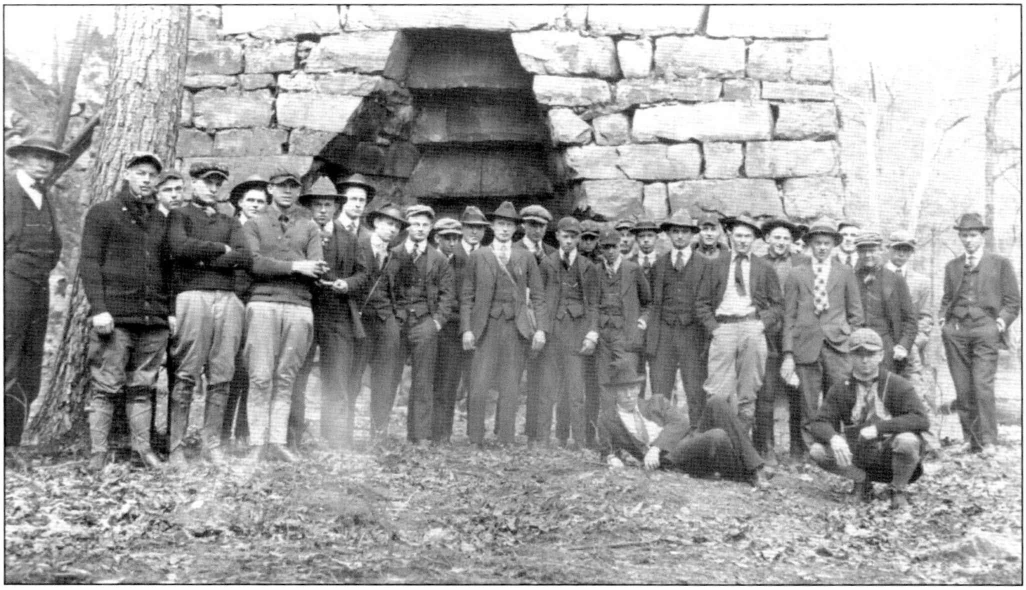

Field trips to the Tannehill Furnace site by geology classes from the University of Alabama were common occurrences in the World War I era, including this visit by students and professors in 1918. Students not only studied the old ironworks but also the area's mineral resources, iron ore, and limestone. (Alabama Department of Archives & History.)

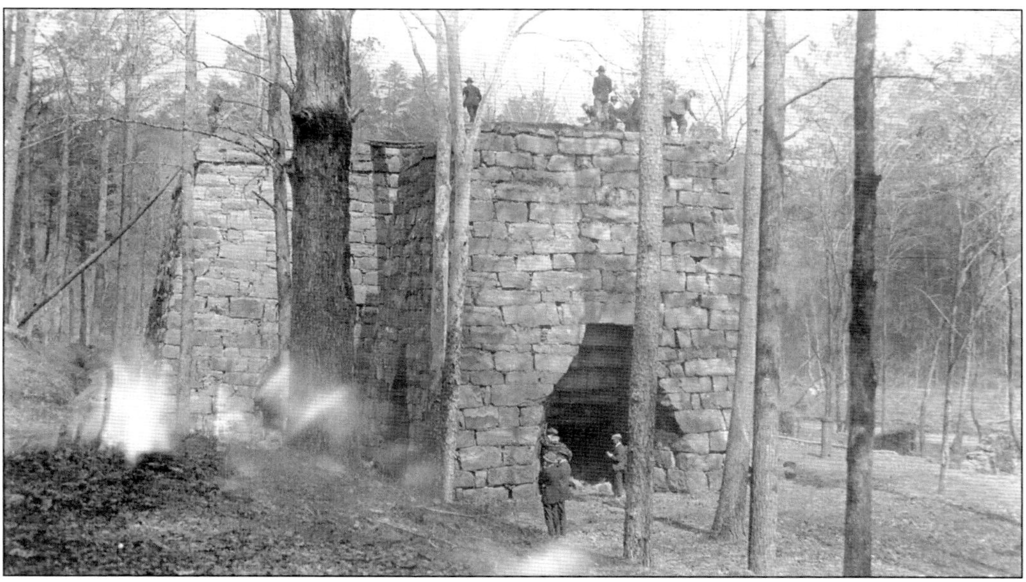

University of Alabama geology students closely examine the remains of the three Tannehill furnaces from top to bottom in this 1918 photograph. Note the makeshift ladder to the left and smoking fires they set to clear the area near the place where the suspected cupola or commissary was located. The brick draft stack can faintly be seen still in place on Furnace No. 2. (Alabama Department of Archives & History.)

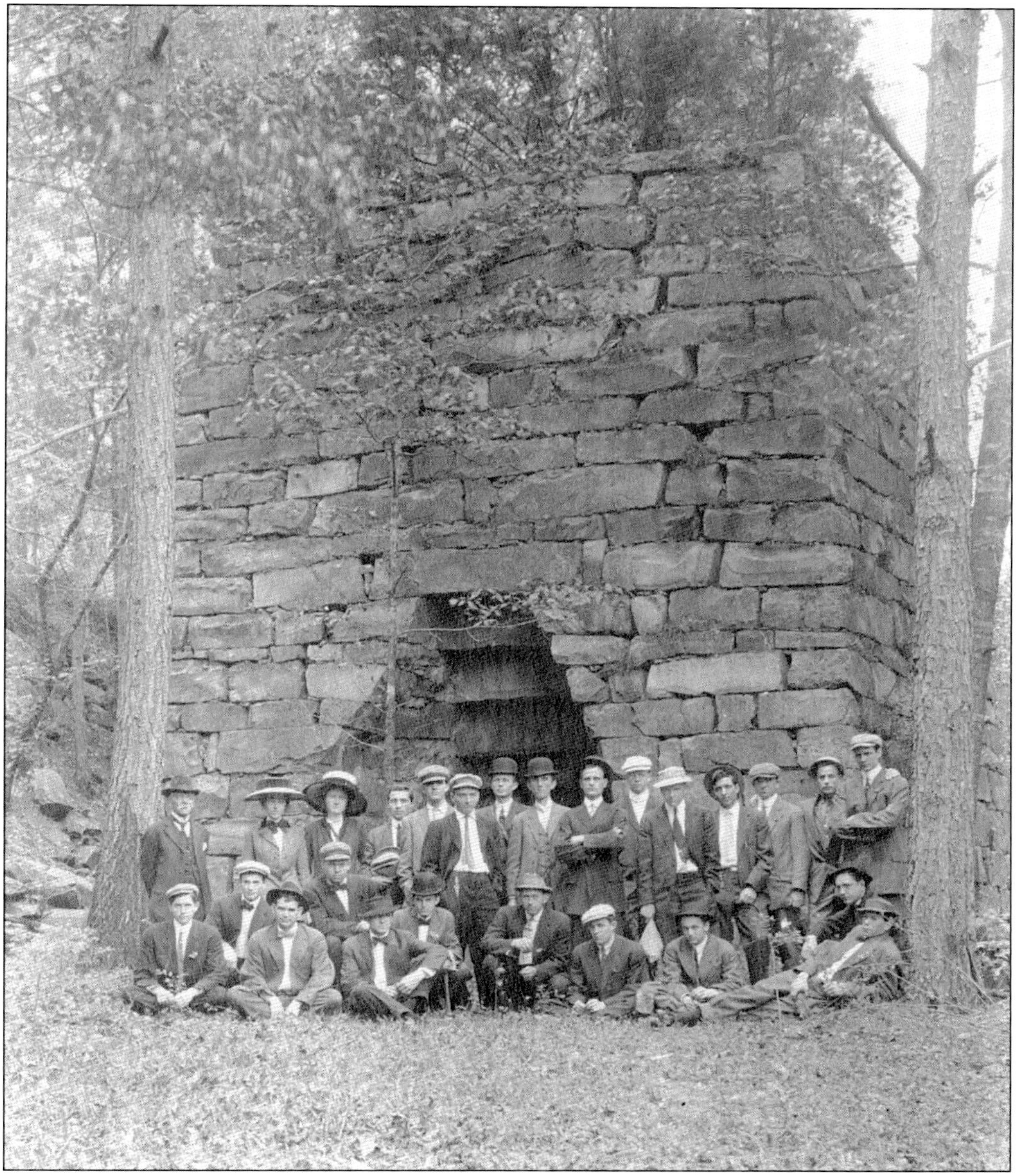

This group of University of Alabama students, likely a geology class, was visiting the Tannehill Furnace ruins around 1905 with state geologist Dr. Eugene Allen Smith (far left, second row). Trees sprout from the furnace yard, which once held sand beds for casting pig iron ingots. Dr. Smith—for whom the current home of the Alabama Museum of Natural History, Smith Hall, is named—was also on the teaching faculty. He was appointed state geologist in 1873 and spent nearly 40 years surveying, mapping, and collecting scientific specimens throughout the state. The cornerstone for Smith Hall was laid on May 28, 1907, construction was completed in the fall of 1909, and it was formally dedicated on May 5, 1910. As state geologist for more than half a century, Dr. Smith cleared the way for the development and utilization of the state's natural resources, discovering, mapping, and publicizing the rich mineral deposits in Alabama. (Alabama Department of Archives & History.)

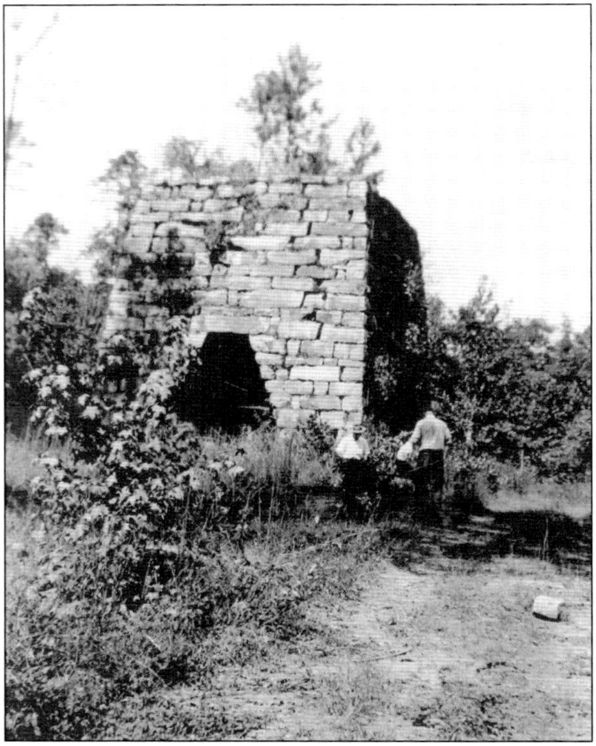

Foundry men visit the Tannehill Furnace site in the 1950s, when examination by *Modern Casting Magazine* showed high arsenic and phosphorus content typical of iron produced from a charcoal and coal fuel mix. Previously, it was thought only charcoal was used as a fuel source at Tannehill. (*Modern Casting Magazine*.)

In 1968, when local Civitan Clubs were pushing the state legislature to declare the furnace site a state historical site and park, unknown vandals dynamited the southeastern corner of the single furnace built in 1859, hoping to dissuade interest in the park movement. The damage only heightened local determination to save the site. No arrests were ever made in the case.

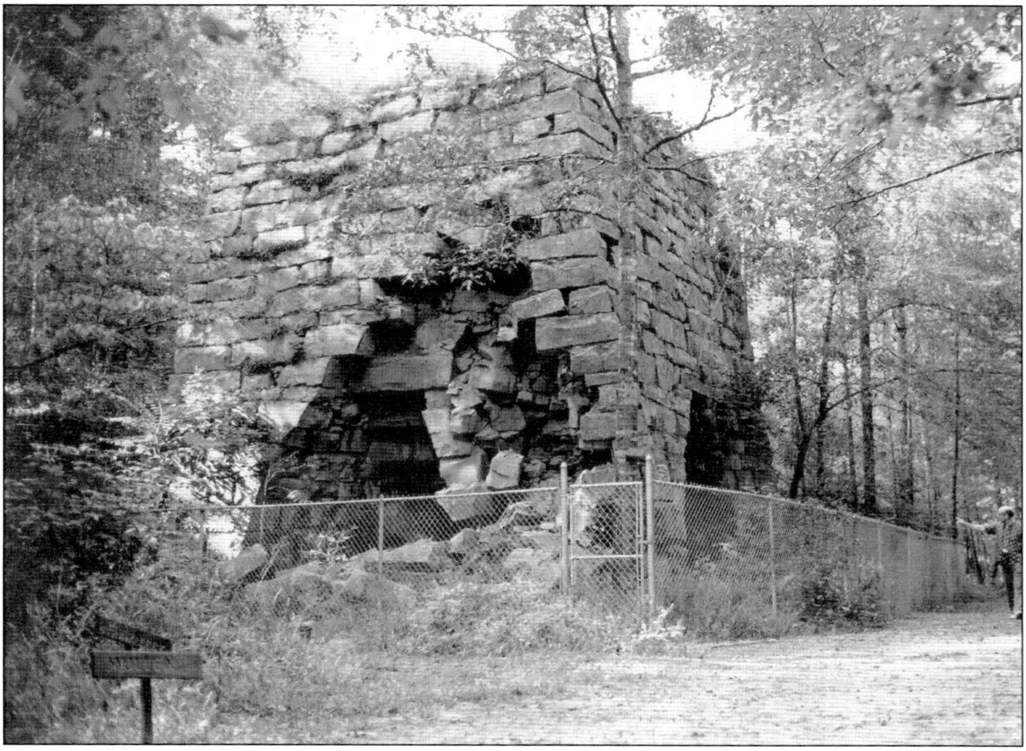

A view from the top of Furnace No. 1 shows the refractory brick lining repairs to the interior core made during its restoration in 1975. The lining on charcoal-fired ironworks of the 1860s had to be replaced periodically during their operation. Eventually, the brick lining began to crumble from the extreme heat, and the furnace was shut down for maintenance and relining. The work involved places hidden from view since the Civil War.

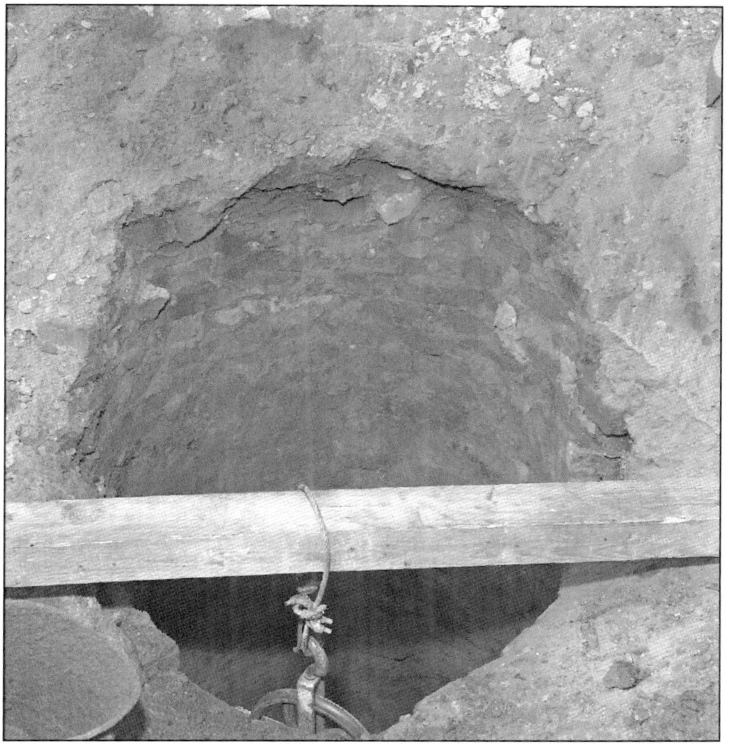

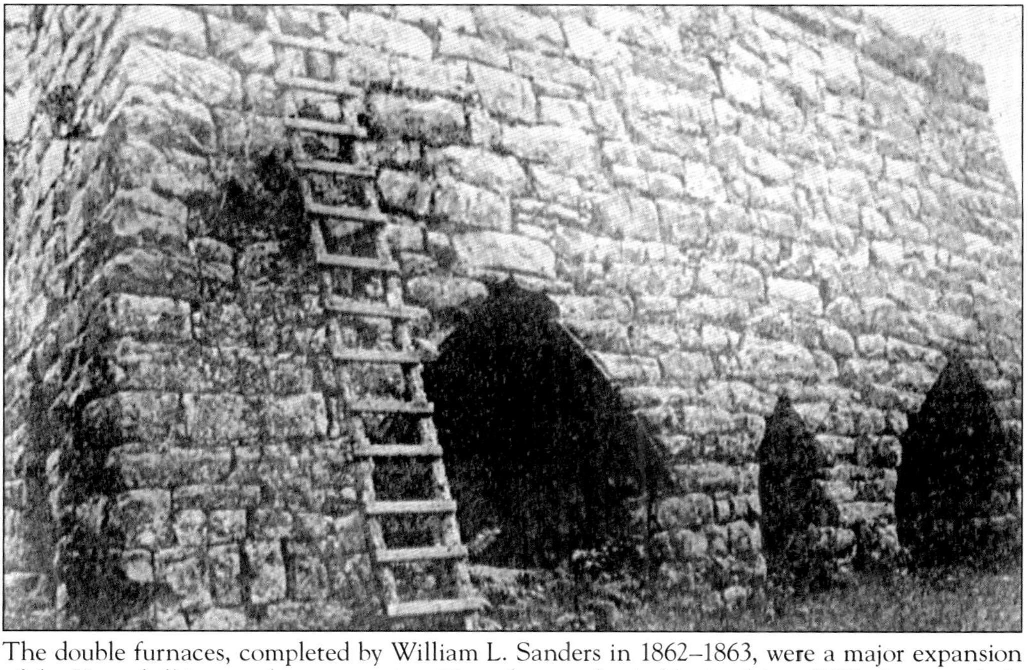

The double furnaces, completed by William L. Sanders in 1862–1863, were a major expansion of the Tannehill iron-making operation. Note the wooden ladder in this c. 1900 photograph. It may have been left by people scavenging for bricks to build houses from the draft stacks atop the furnaces. (Scenic South Card Company, Bessemer, Alabama.)

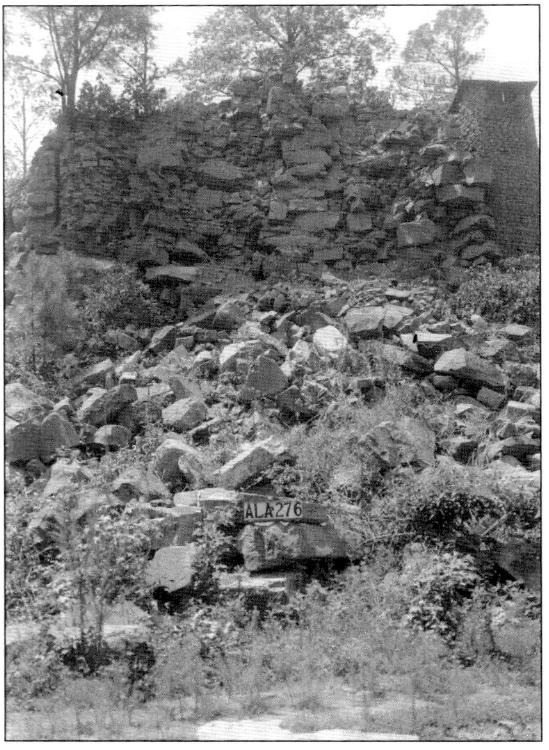

The double furnaces at Tannehill, which were built during the Civil War, fell under their own weight in March 1930 and remained in a state of disrepair until efforts to restore them began in 1985. One of the most extensive ironworks restoration projects in the nation, the park invested over $112,000 in putting this part of the furnace battery back together. (Alex Bush, HABS photograph AL-276-4, National Park Service.)

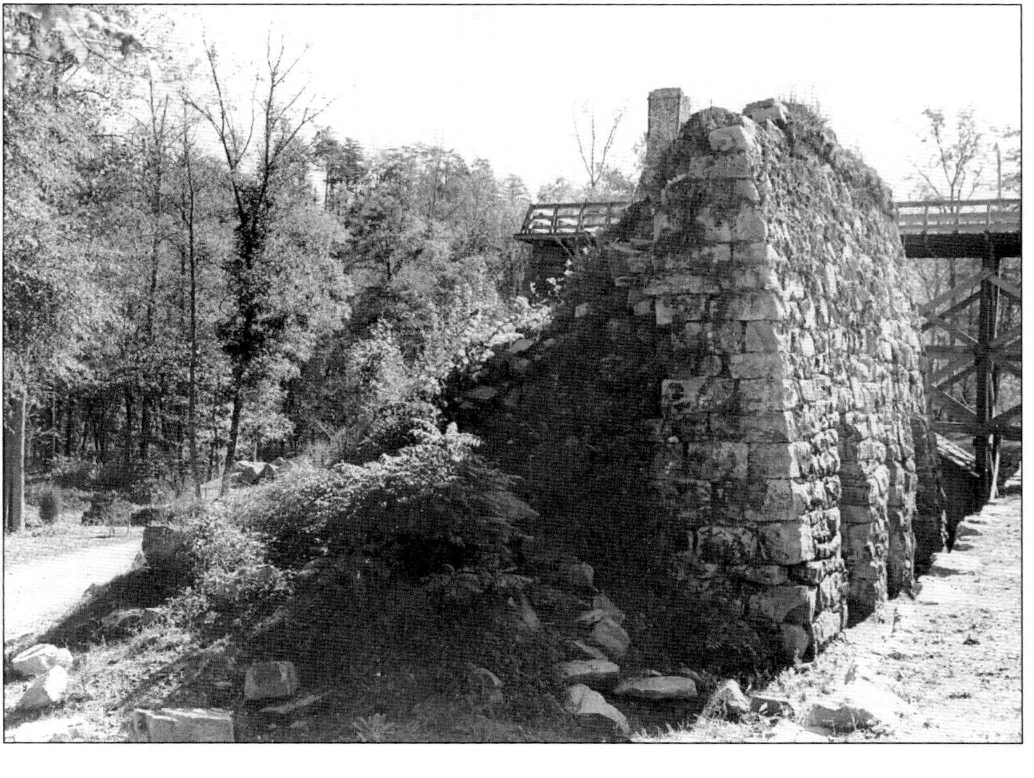

Having started under the supervision of Moses Stroup in 1862, the double furnaces were completed a year later by William L. Sanders. Sanders was a Selma merchant who had entered into an agreement with John Alexander to buy the Roupes Valley Ironworks at Tannehill during the war. The double furnaces fell in 1930, some 68 years after they were erected, due to ground subsidence and effects of weather.

This 1936 photograph shows the old flywheel looking south toward the fallen facade of the double furnaces. Note the huge iron plate at the top of Furnace No. 3; it was probably used as the draft stack base. The inner-core bricks are clearly seen after the collapse of 1930. (Alex Bush, HABS photograph AL-276-2, National Park Service.)

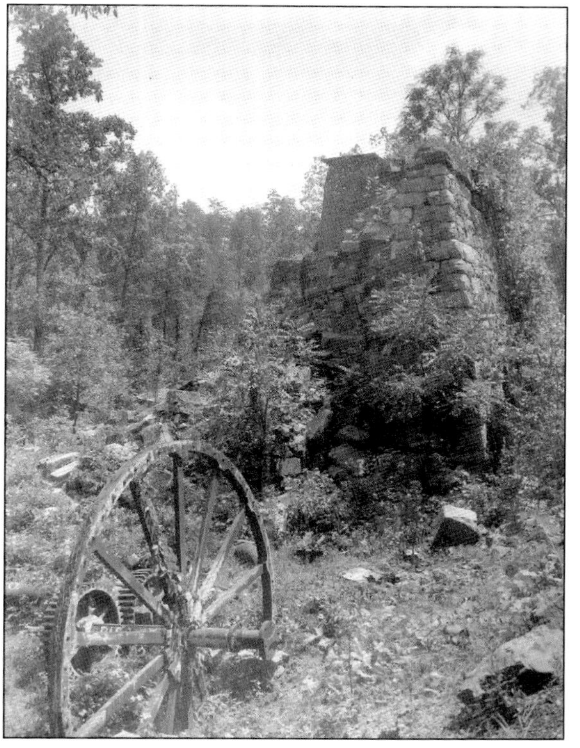

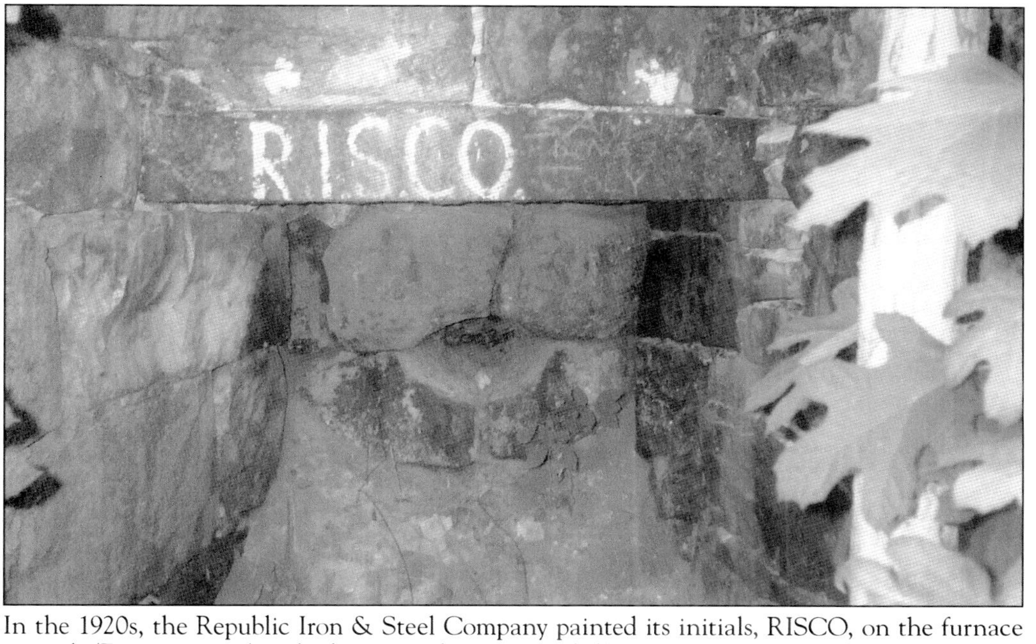

In the 1920s, the Republic Iron & Steel Company painted its initials, RISCO, on the furnace mantels (Furnace No. 2) and other iron relics at Tannehill in order to discourage metal scavengers from hauling off tons of scrap iron from the historic furnace yard. If anybody was going to make use of the scrap metal, it was going to be the company. Tons of artifacts were later thrown into its furnaces at Thomas.

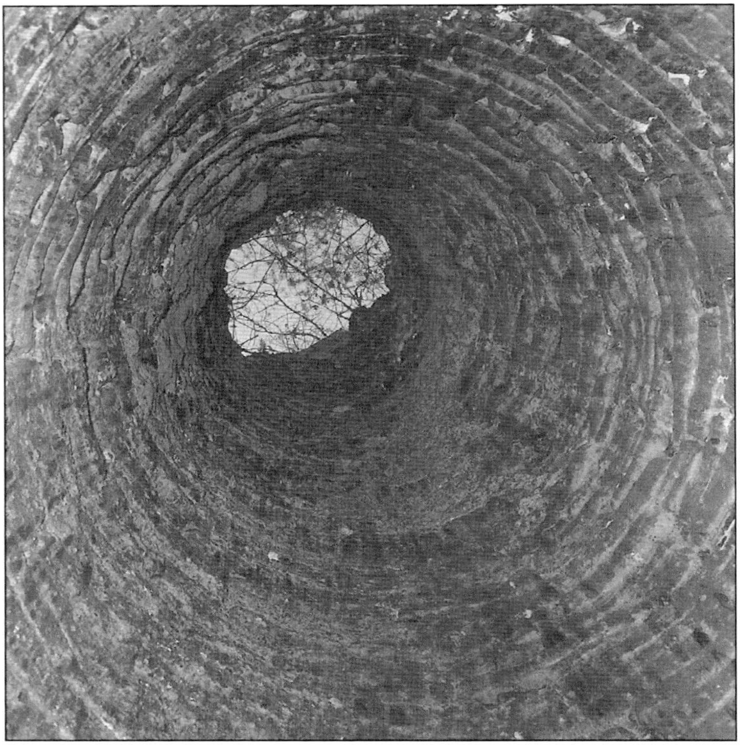

The inner core of Furnace No. 1 in 1936 revealed a glassy appearance from the extreme heat of the 2,600-to-3,100-degree charcoal fires that smelted iron ore into molten iron beginning in 1859. The chambers were lined with fire brick to withstand the high temperatures it took to make metal. (Historic American Buildings Survey, AL-246-6, National Park Service.)

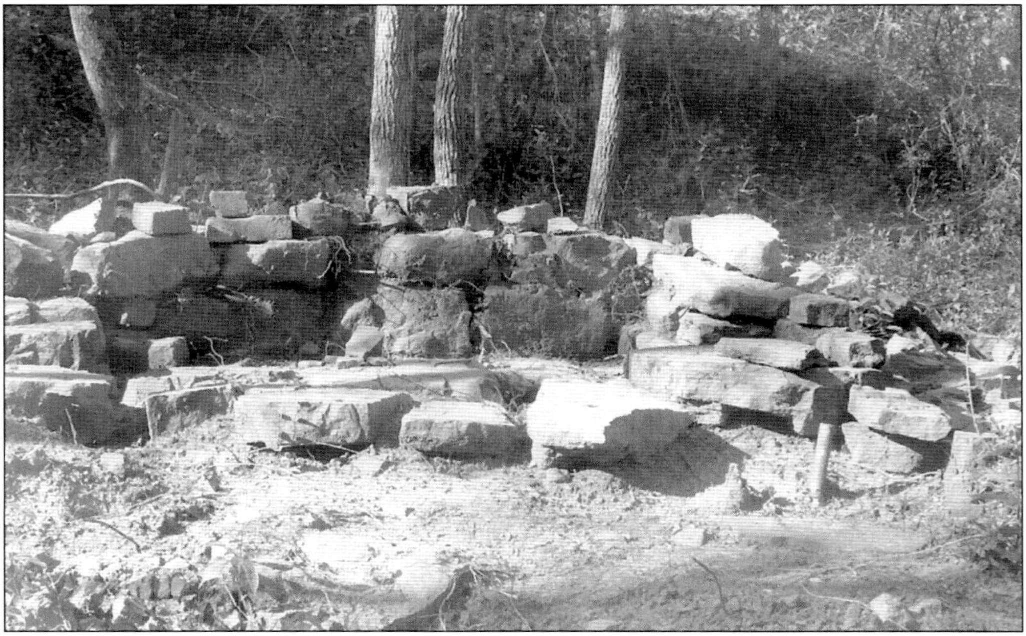

Excavations early in 1986 revealed remnants of a building behind the furnaces, which was probably used as a commissary or cooking area to feed shift workers. Among artifacts recovered was a "Number 4 Oven" manufactured in the foundry. This oven was most likely used to prepare meals. A second fireplace is a short distance away.

The south furnace crucible for Stroup's 1859 furnace is shown here as it appeared in 1936. During the Civil War, hundreds of tons of pig iron poured from this arch. The pig iron was then shipped to the Naval Gun Works at Selma for all the weapons of war. This furnace made use of a waterwheel for air blast. (Alex Bush photograph, HABS No. AL-276-5, National Park Service.)

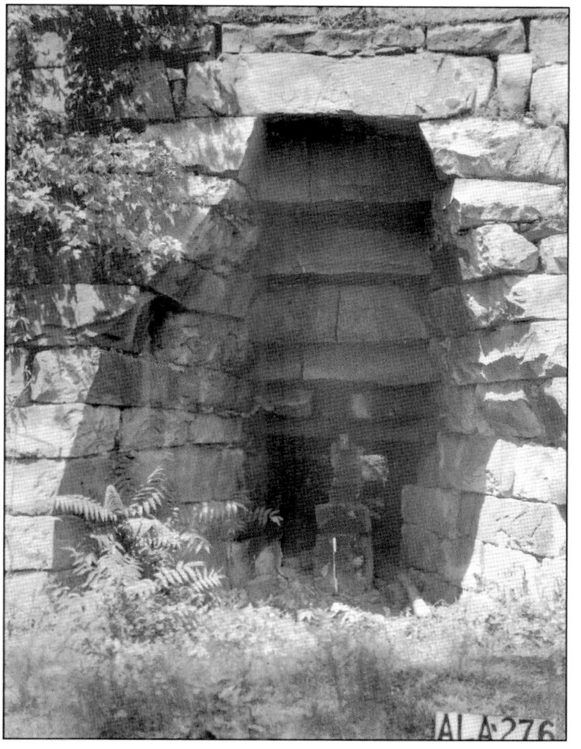

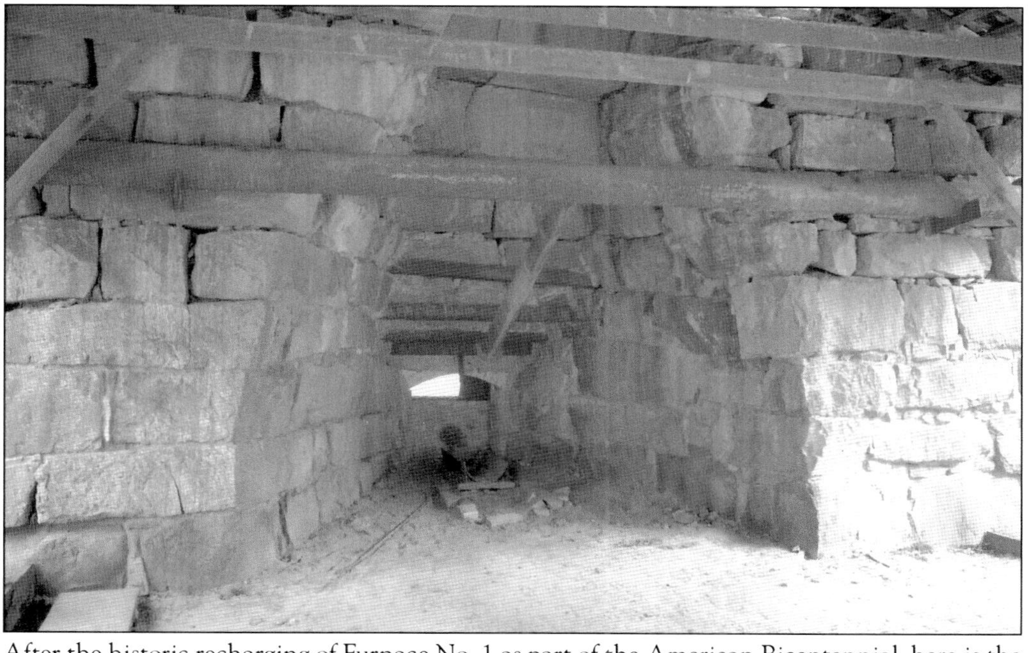

After the historic recharging of Furnace No. 1 as part of the American Bicentennial, here is the same location as it appeared in 1977. The air blast pipe connected to an industrial blower was added to expedite the metal melting process. A tamping rod, which was used to break up the clay plug to allow the flow from the furnace, still rests to the left side of the picture.

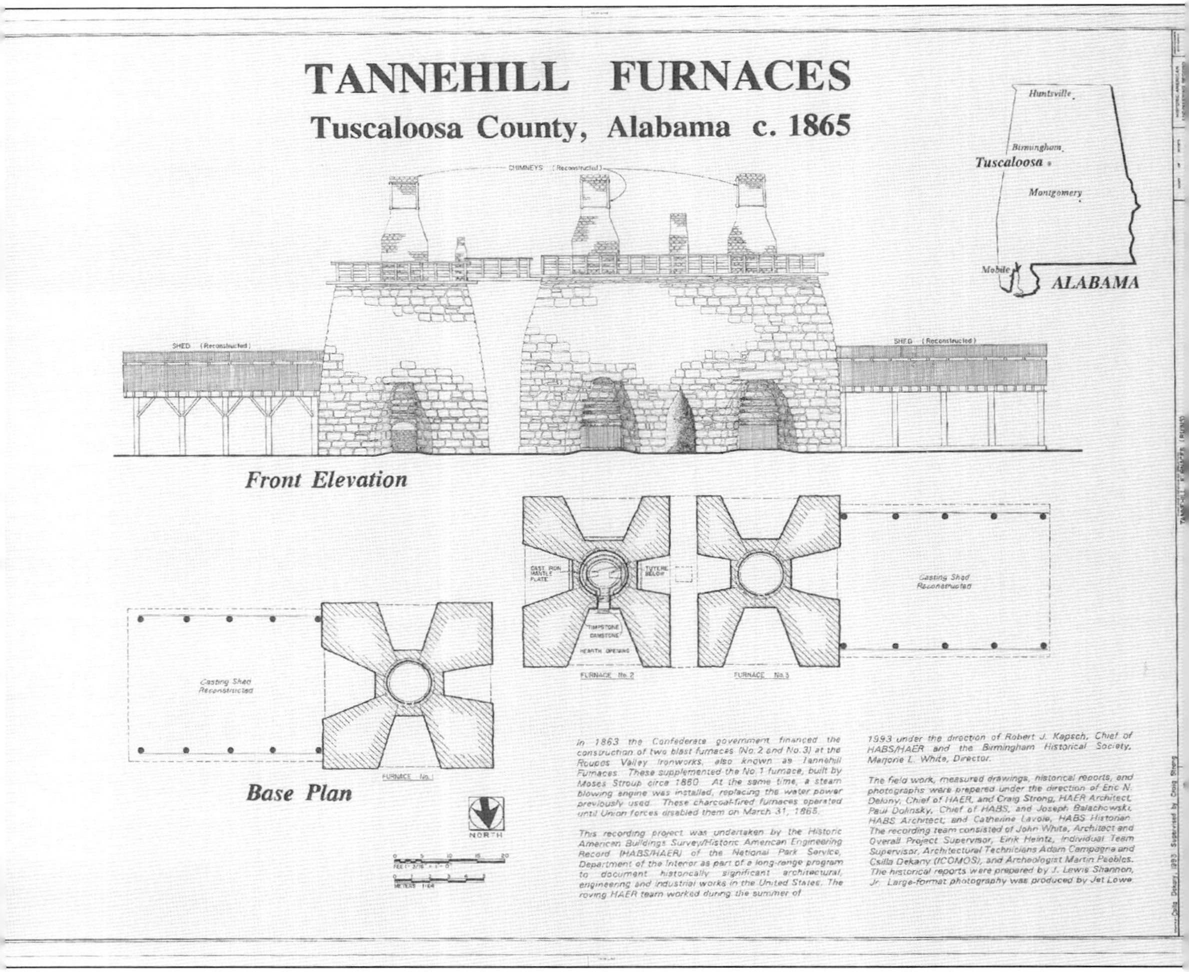

In 1993, the Historic American Buildings Survey (HABS) and the Historical American Engineering Record (HAER), both a part of the National Park Service, drew up this base plan and front elevation of the Tannehill Furnaces, which mirrored the furnaces' appearances from around 1865, as part of ongoing plans to develop the Birmingham Industrial Heritage District. The recording project was aimed at documenting historically significant architectural, engineering, and industrial works in the Birmingham area. (Robert J. Kapsch, HABS/HAER, National Park Service.)

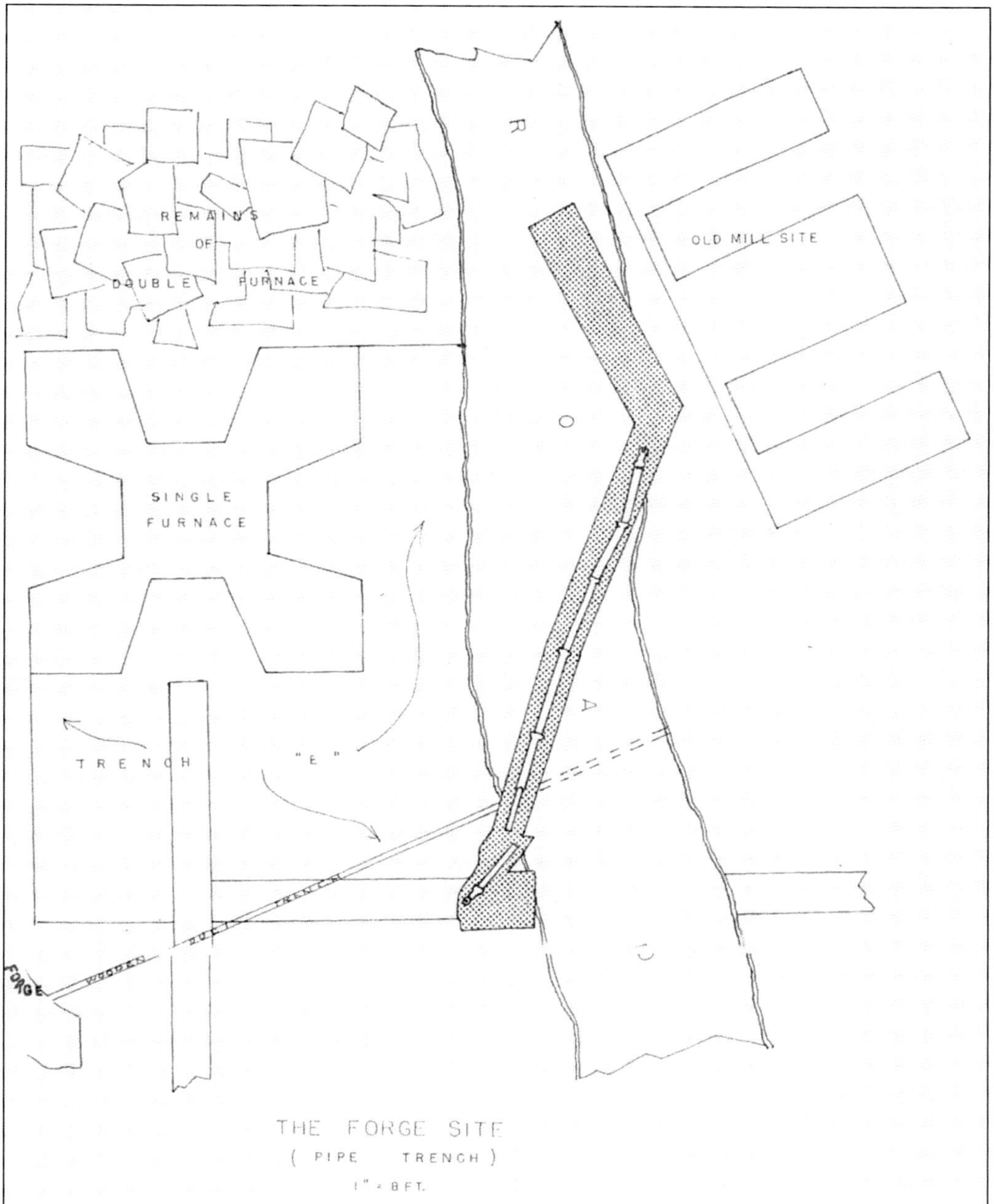

This archaeological drawing of the forge site at the Tannehill Ironworks was prepared by the University of Alabama in 1956 as part of its field report documenting features of the furnace complex. The work located Hillman's original 1830 forge near the present-day blower house for Furnace No. 1. (David L. and Thomas H. DeJarnette, *Archaeological Investigation of the Tannehill Blast Furnace*, Alabama Museum of Natural History.)

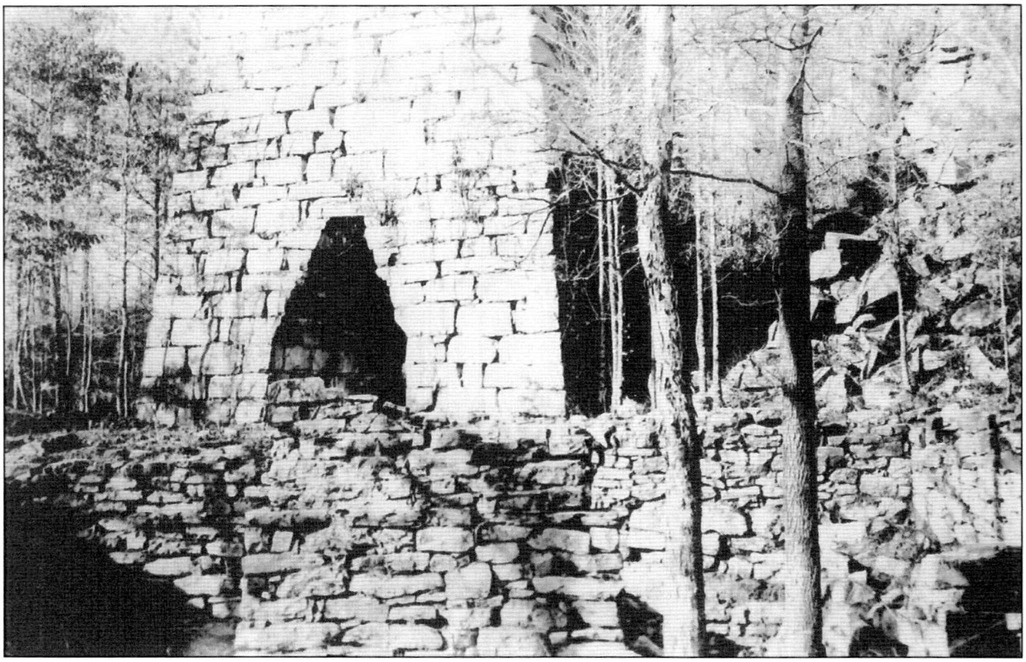

This view of the original furnace at Tannehill was made a part of the University of Alabama archaeological study that noted the key remaining features of the blast furnace site. It was taken from the remains of the blower house, which was the location of a waterwheel that provided air blast and perhaps ran a machine shop. (University of Alabama.)

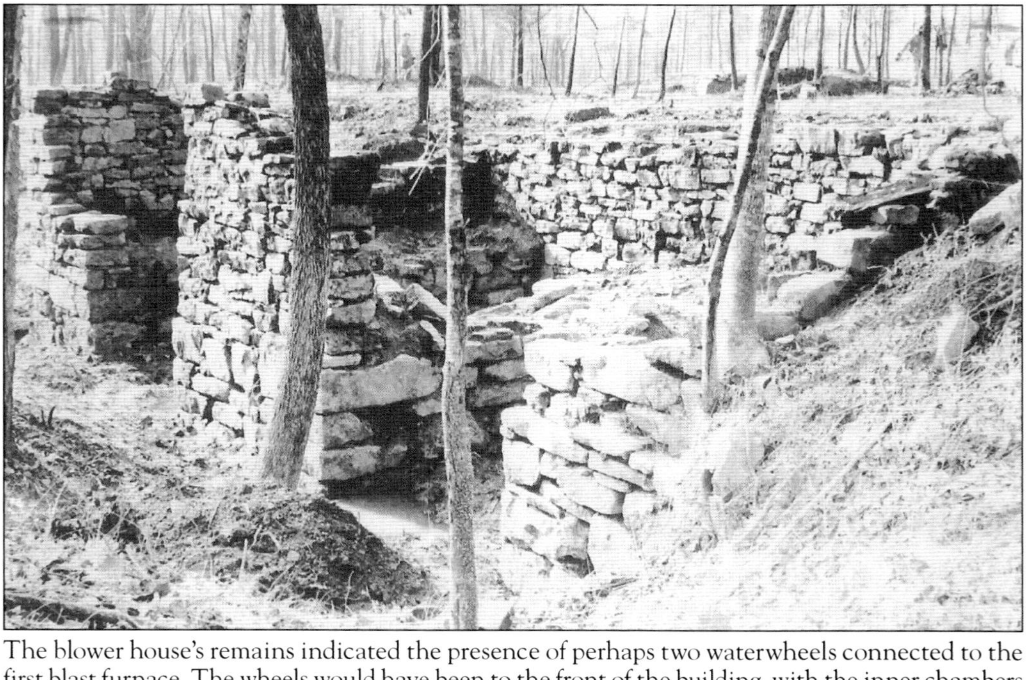

The blower house's remains indicated the presence of perhaps two waterwheels connected to the first blast furnace. The wheels would have been to the front of the building, with the inner chambers holding crank arms, which powered blowers and machine gearing. (University of Alabama.)

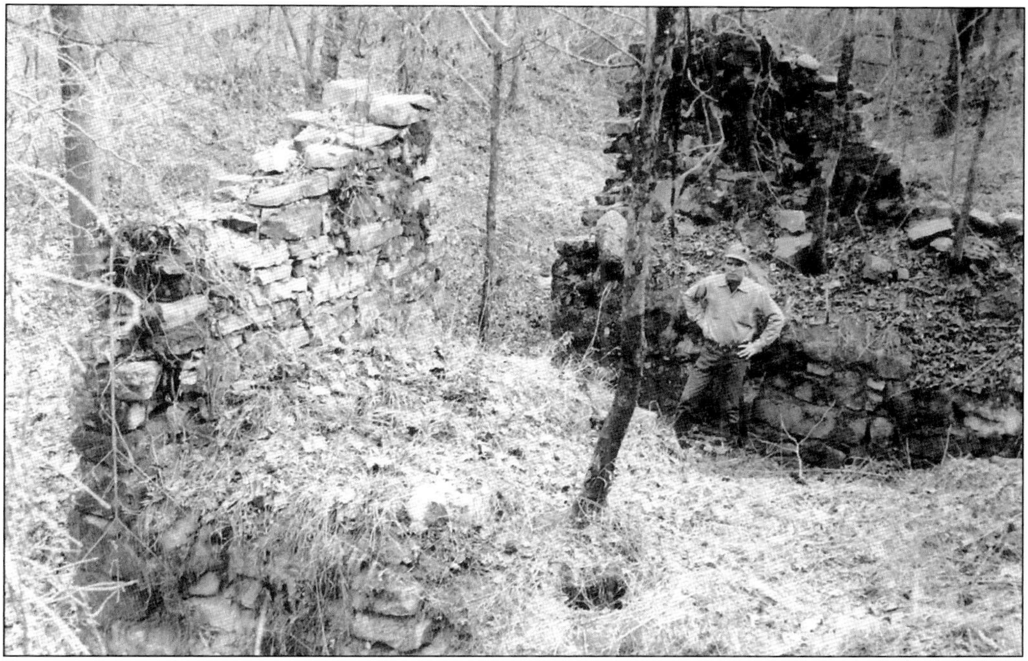

One of the two principal investigators during the University of Alabama's 1956 archaeological study of the Tannehill blast furnace site, David DeJarnette, stands inside the remains of the old blower house. The study lasted three months, from January 17 to March 20. (University of Alabama.)

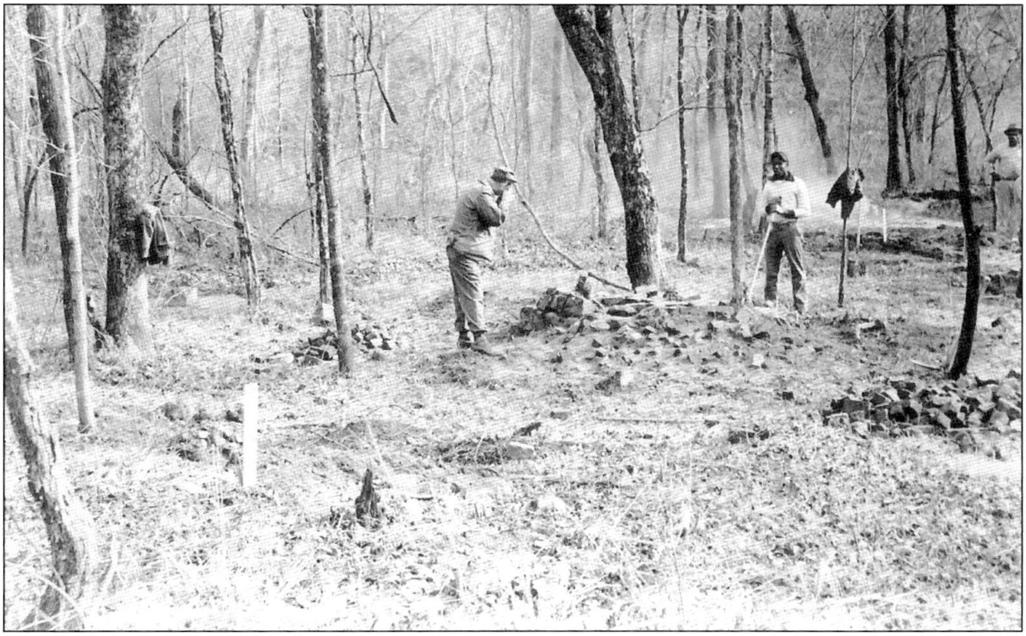

University of Alabama archaeologists map the office location of the Tannehill works during their archaeology study of the site. Located here was a two-room structure with a central fireplace opening into each room. Most likely, company records were kept in this building. First thought to be a nail factory, it was burned by Federal forces on March 31, 1865. (University of Alabama.)

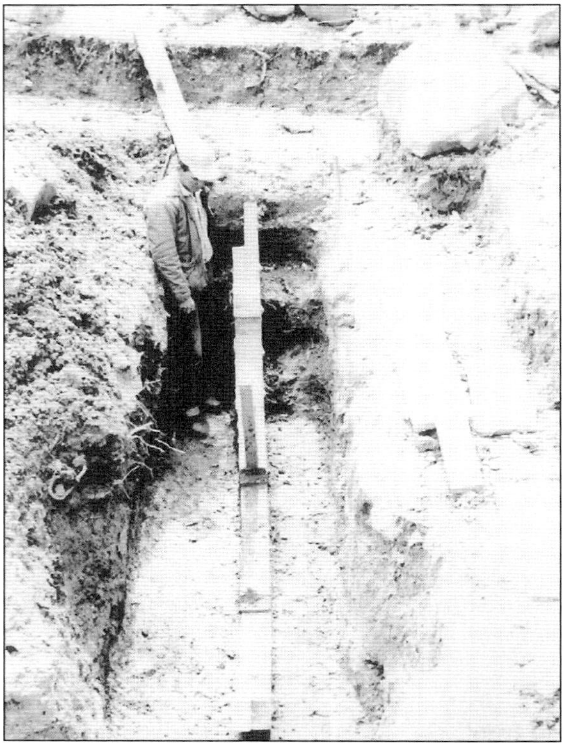

The steam engine's flywheel in the steam house measured 14.5 feet across with spokes spaced at three-foot intervals. It was so large that a part of the wheel turned in a trench, the location of which ultimately saved its bottom portion from metal scavengers. The bolted segmented sections are thought to be still buried near Furnace No. 3, where archaeologists left the find in 1956. (University of Alabama.)

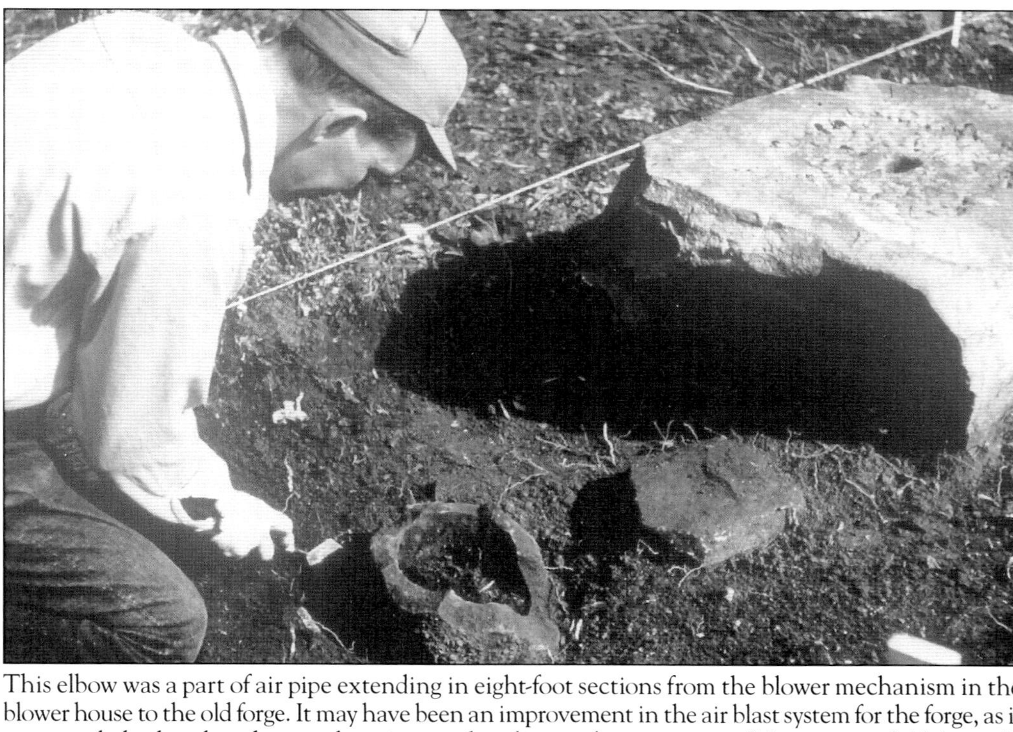

This elbow was a part of air pipe extending in eight-foot sections from the blower mechanism in the blower house to the old forge. It may have been an improvement in the air blast system for the forge, as it apparently had replaced a wooden pipe used earlier in the same area. (University of Alabama.)

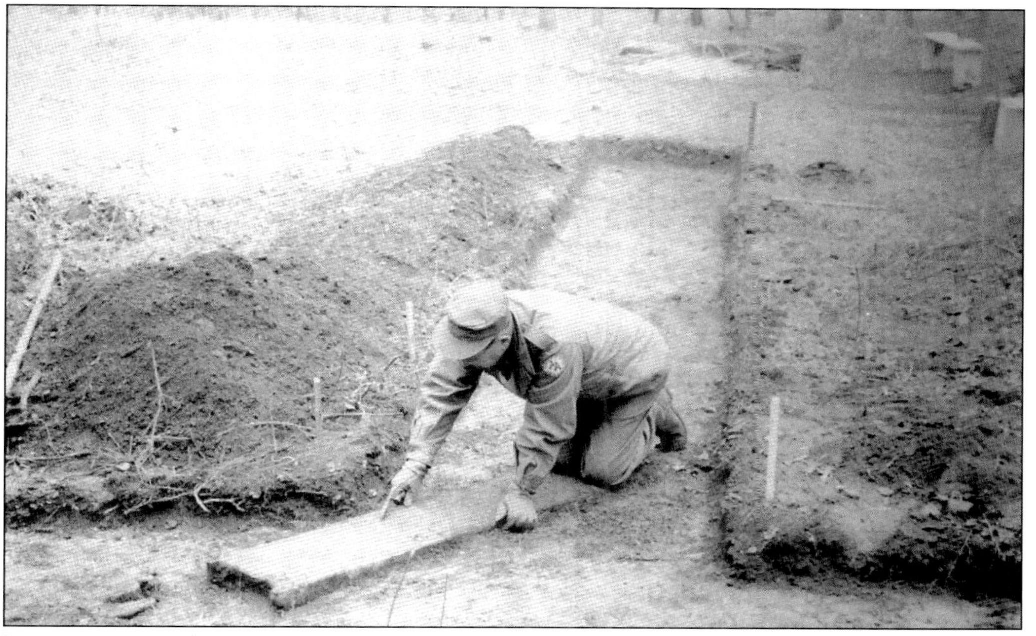

Large amounts of iron castings, machine parts, iron fragments, and casting pieces were found in the trench cuts during the excavation, including this large slab that may have had either a structural use or served as iron flooring. Many of the objects are either on display at the Iron & Steel Museum of Alabama or in storage. (University of Alabama.)

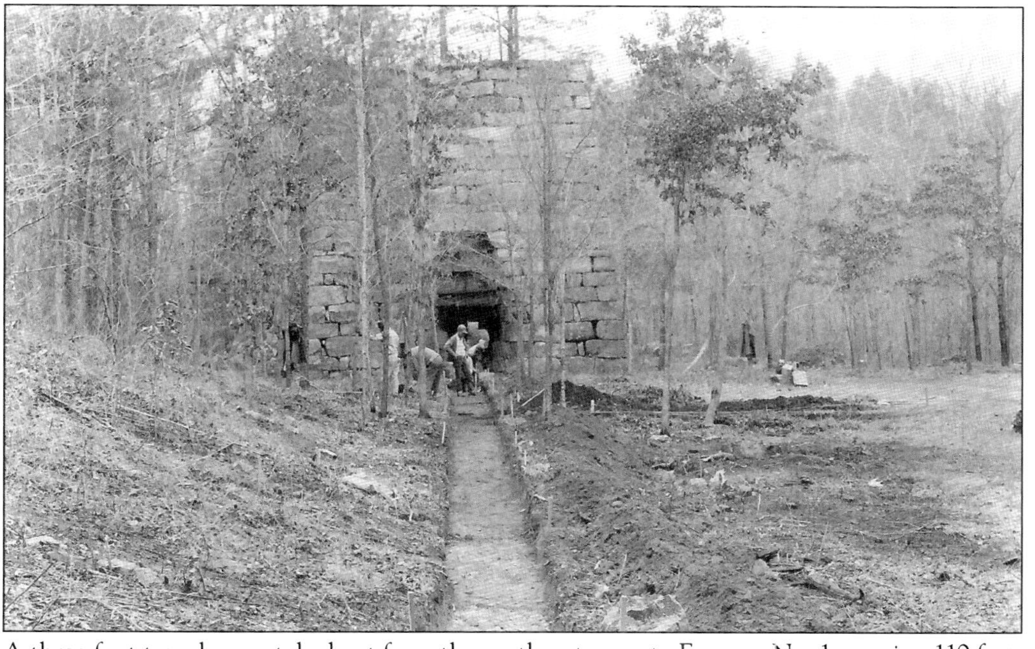

A three-foot trench was staked out from the south entrance to Furnace No. 1 running 110 feet. This trench was excavated at three-inch levels, and a tremendous amount of iron and casting pieces were found. The wooden duct ran into this area heading toward the old forge site, just to the south of where the first blast furnace would later be built. (University of Alabama.)

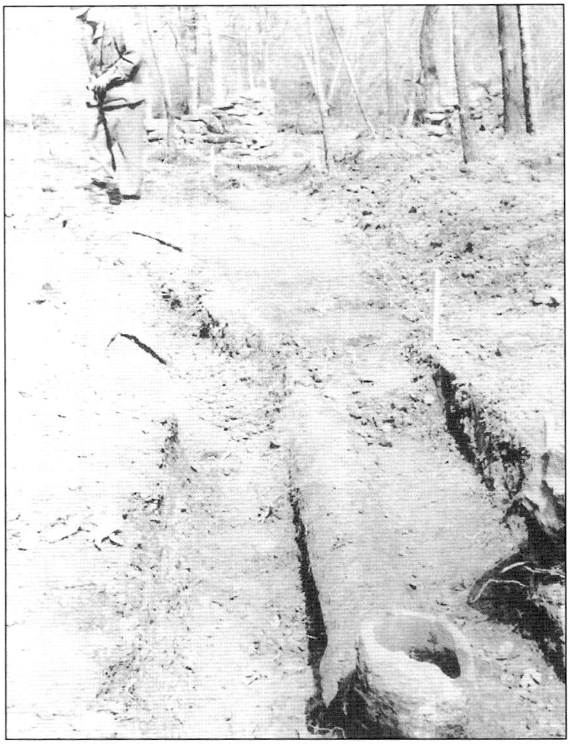

The elbow with further excavation begins to show the extent of the air blast pipe system, which took air blown by tubs or a bellows to the forge to enhance the temperature of the forge fires. Hillman's bloomery had two fire boxes. The apparatus probably dates to the early 1840s–1850s. (University of Alabama.)

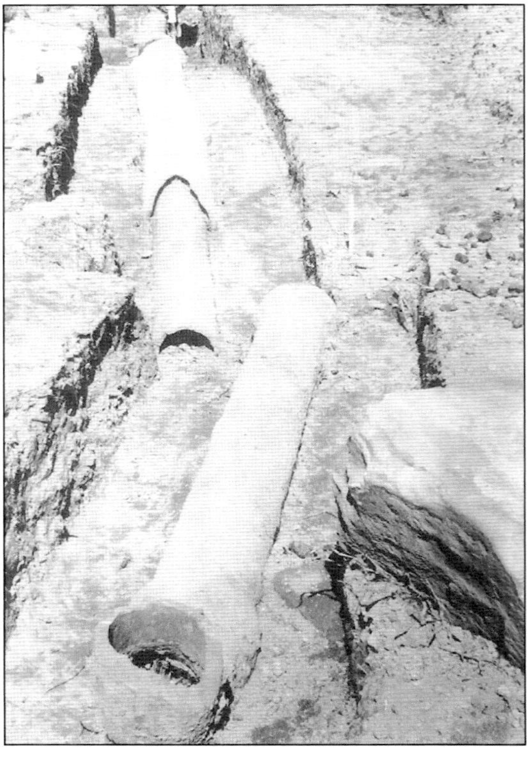

Here, three sections of the heavy 12-inch cast iron pipe lie in the blower trench leading from the creek to the forge. The pipe was in remarkably good shape when excavated in 1956 and is now on display in the Iron & Steel Museum of Alabama. The iron pipe replaced an earlier wooden duct running in the same direction in 1830. (University of Alabama.)

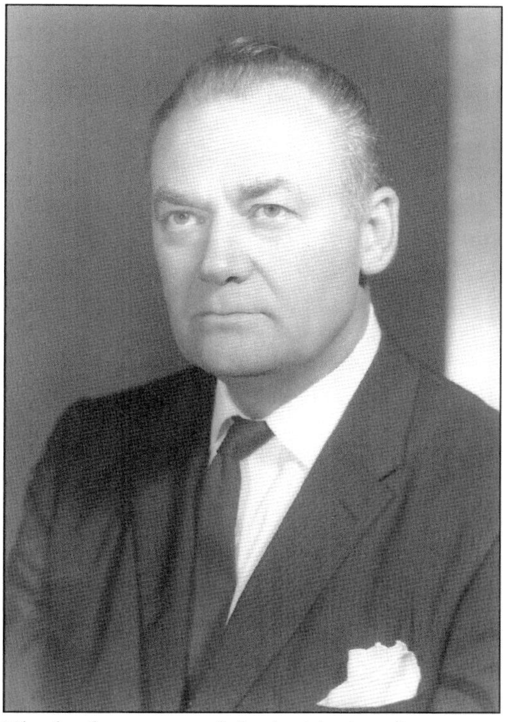

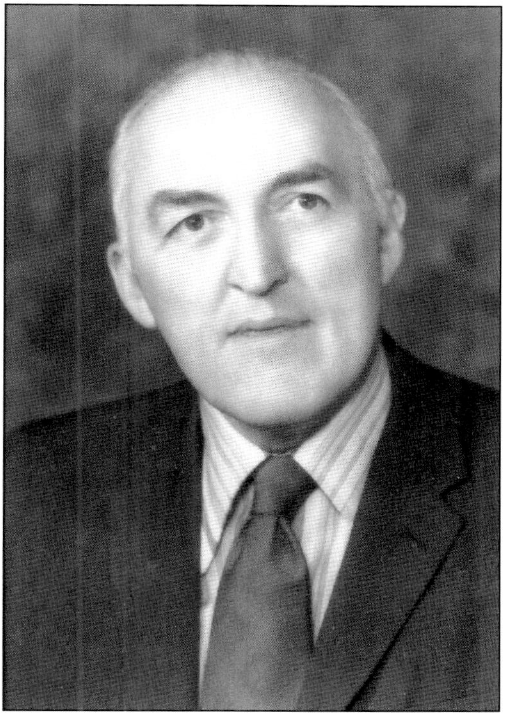

The lead sponsors of the legislation that created Tannehill Ironworks Historical State Park in 1969 were, from left to right, Sen. E.W. Skidmore of Tuscaloosa and Sen. Eddie Hubert Gilmore of Bucksville. The bill, SB 748, was approved on September 12, 1969, and called for the preservation, restoration, maintenance, and promotion of the site as a state park and historic site. The park's first manager was Ed Nelson, the former manager of the Holiday Beach Resort at Woodstock and also a member of the Woodstock Civitan Club, which had been instrumental in promotion of the park concept at Tannehill. Nelson, who began to add period buildings to the park, would serve until his death in 1991.

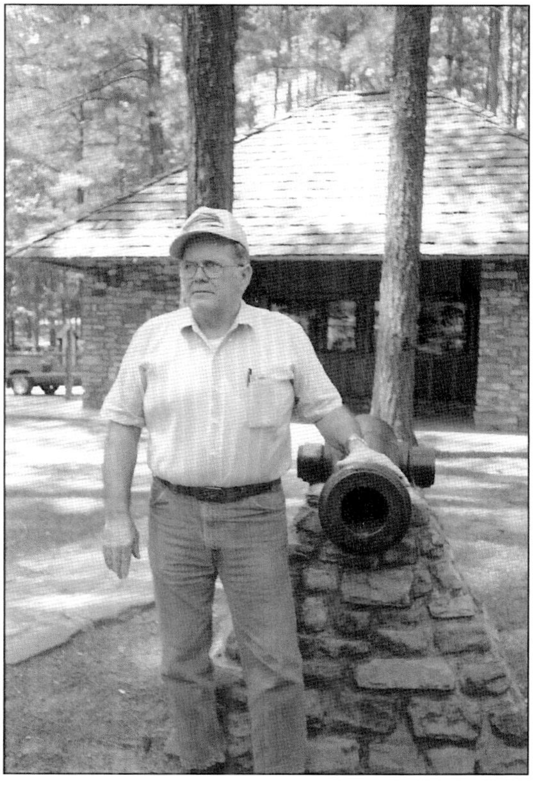

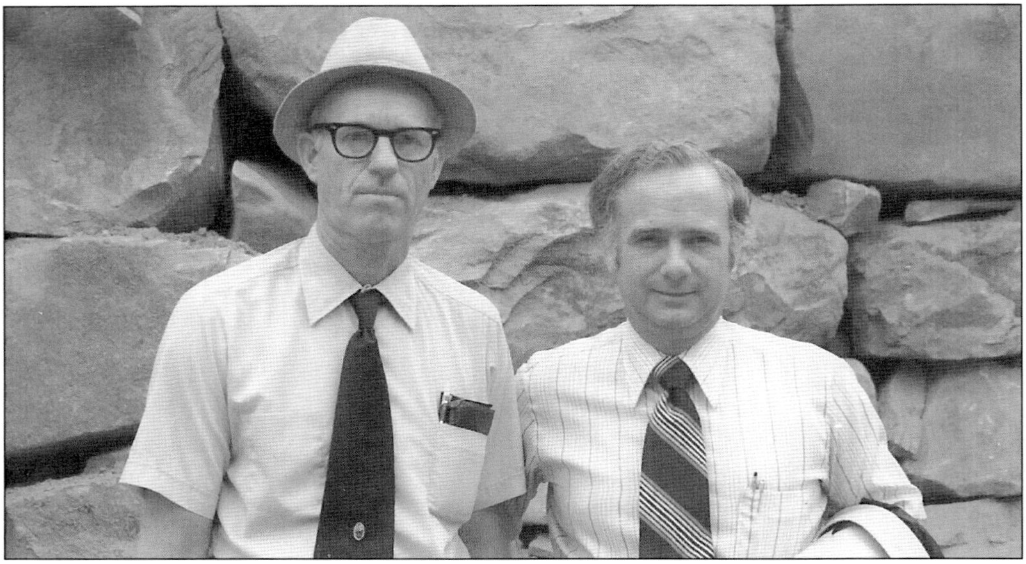

Two of the earliest supporters of a state park at the Tannehill ruins were Ed Lewis (left), a US Steel employee who served as lieutenant governor of the Alabama Central District for Civitan International in 1967, and state representative Walter Owens of Centreville, who pushed the park legislation through the House in 1969. They are shown at the furnace site in 1976.

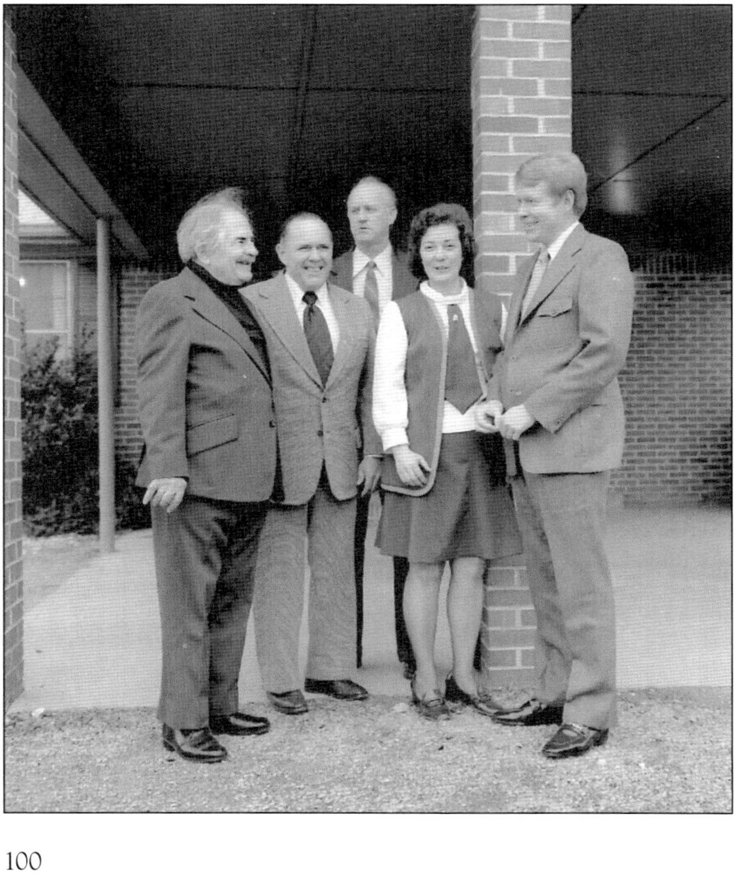

The Tannehill Furnace & Foundry Commission, the state agency created under the legislation to administer the new park development, held an annual meeting at the University of Alabama, which was where a group of professors and administrators first proposed the preservation efforts. Shown in a c. 1975 photograph are, from left to right, Dr. Charles Summersell, chairman of the Southern History Department at the University of Alabama; park manager Ed Nelson; and board members Ed Lewis, Joyce Franks, and Jim Bennett.

The man who rebuilt the fallen front exterior walls of Furnaces Nos. 2 and 3 was Edwin Robinson, a heavy stone mason from Cullman whose craftsmanship closely resembled that of the 19th-century furnace builders. The work took place between 1985 and 1987, repairing the damage caused by the collapse in 1930.

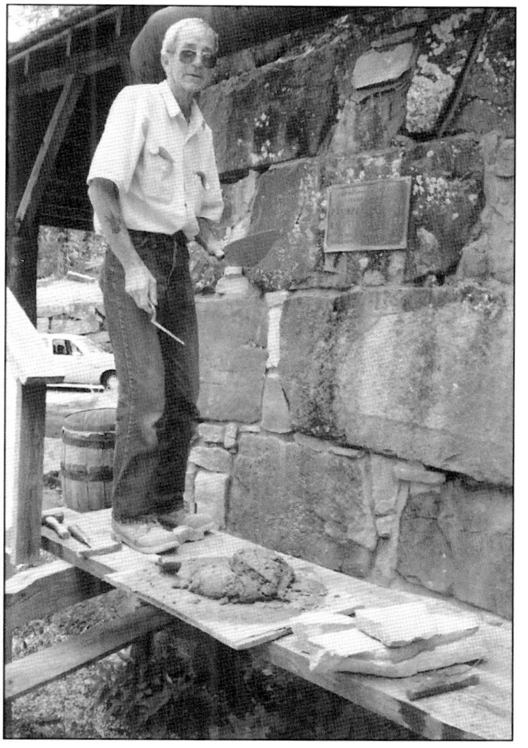

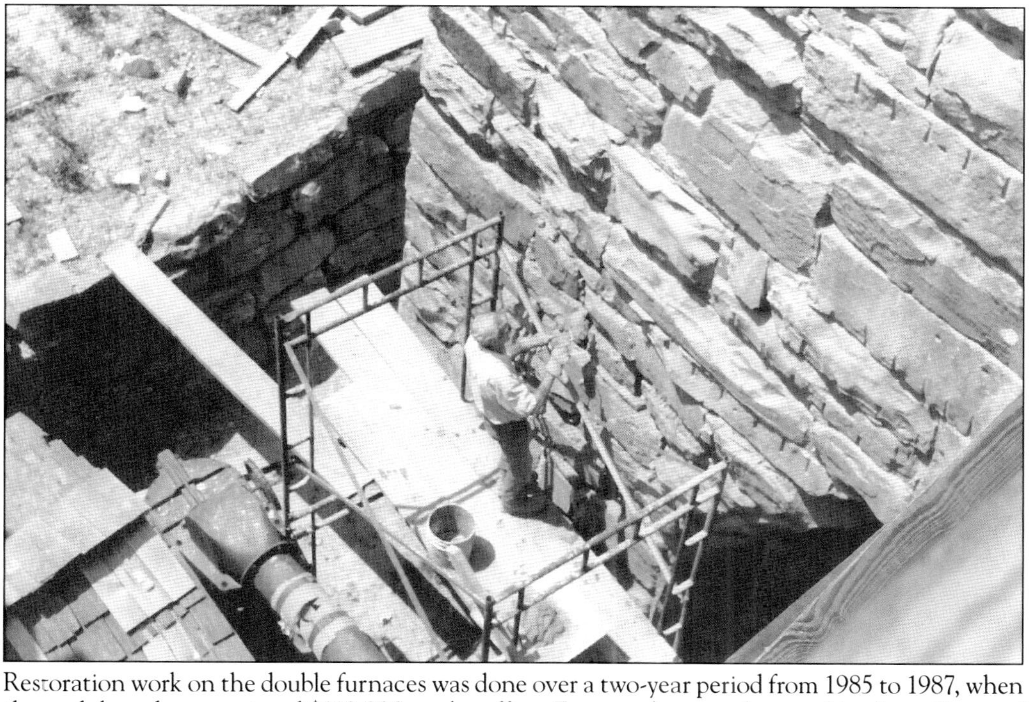

Restoration work on the double furnaces was done over a two-year period from 1985 to 1987, when the park board appropriated $112,000 to the effort. Fractured stones damaged in the collapse of the furnace fronts in 1930 were replaced with similar stones from the Blocton Coke Ovens.

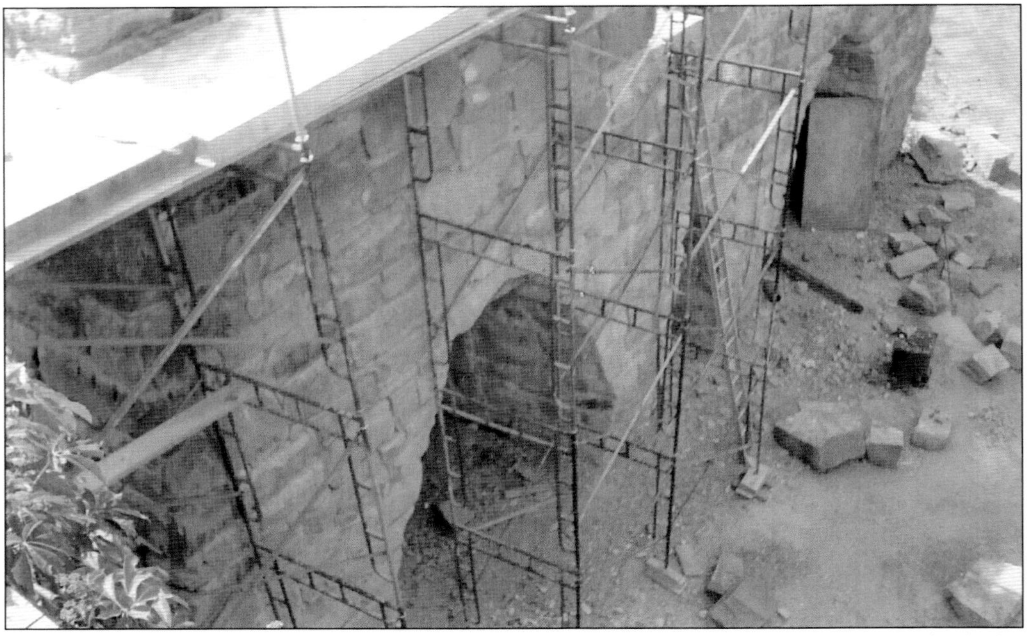

Scaffolds are shown in place during the double furnaces' restoration in 1985, which rebuilt the fallen front portion of the blast furnaces that was originally financed with Confederate bonds in 1862. These works featured hot blast stoves and a steam engine to drive air blast. They were among the most modern in Alabama for their day.

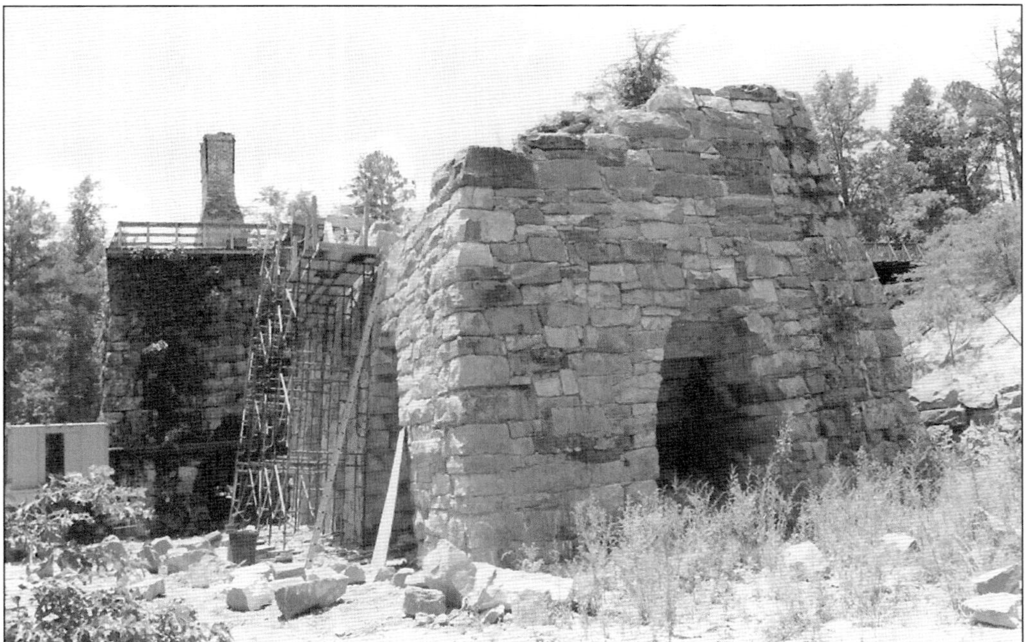

This photograph from 1985 shows work on the north end of Furnace No. 3. Concrete was used in the reconstruction project to make sure the blast furnace stays in place for another 100 years. These works suffered a collapse in 1930 due to ground subsistence, as they were partially built on a slag fill.

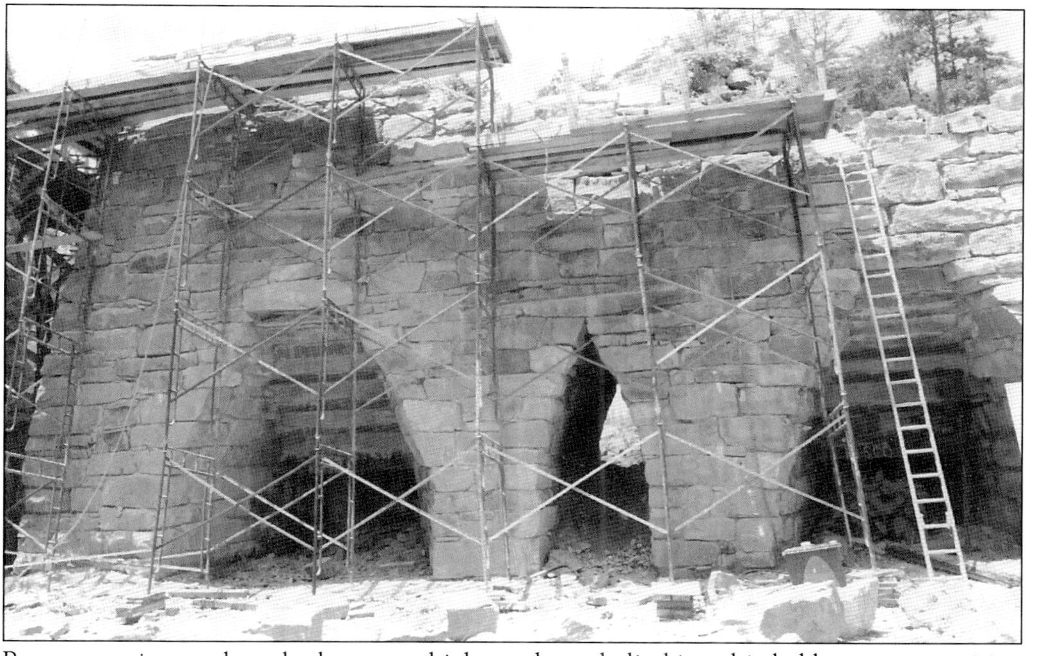

Reconstruction work took place on a high perch, and climbing this ladder was an exciting experience. The furnaces stand 40 feet above the ground. The brick draft stacks added another 10 to 12 feet. The heavy stone face not only included original sandstone blocks cut by slaves but also replacement stones from the Blocton Coke Ovens.

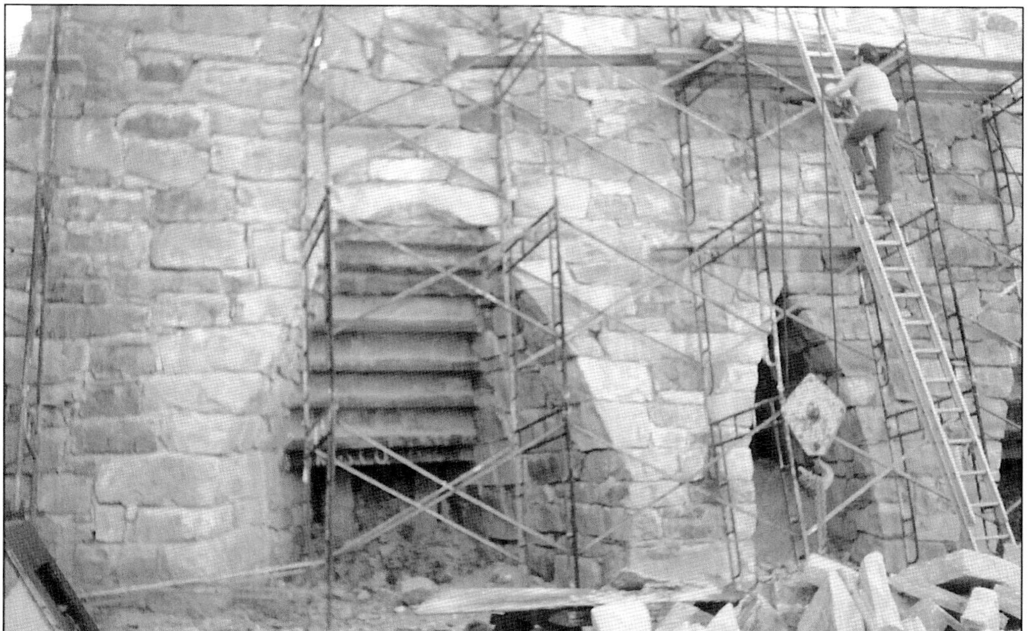

Ladder work was a bit scary, because workmen traveled up and down the furnace face to carry on the repair work and to direct the cranes lifting the heavy stones to the top layers. Great care was taken to make the restoration as historically accurate as possible, including the stone shapes and sizes.

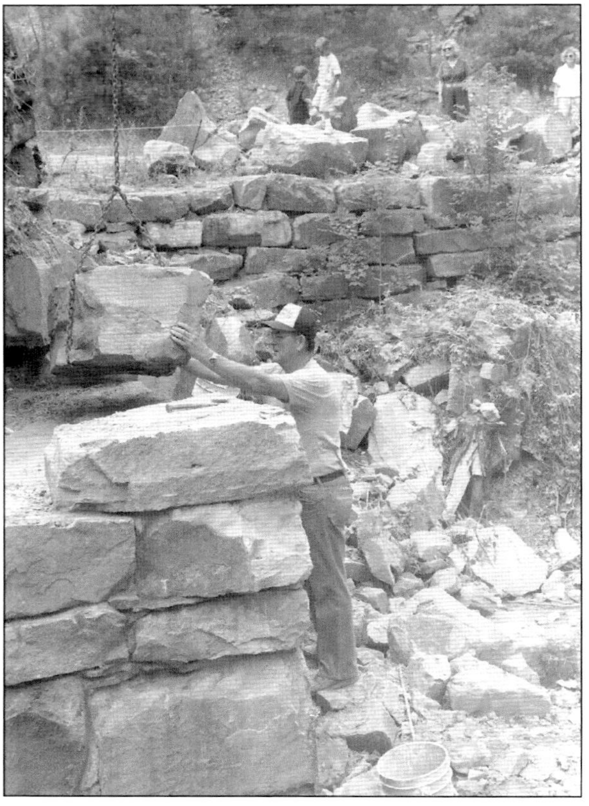

Joe Middlebrooks, a park employee, helped lay the stone work on the furnace fronts. Some of the stones weighed 400 to 600 pounds a piece. US Steel donated some of them from other sites to match what was already put in place by slave stone workers in the 1860s. A sandstone quarry is a short distance from the furnace site.

This 1985 photograph shows the interior core of Furnace No. 2, including charcoal fire burns still in evidence from the intense heat of ore reduction. The brick liner is missing from this furnace, but the liner in Furnace No. 3 was largely in place, which indicates perhaps No. 2 was out of blast the day of the Federal attack. This view is in the forge area.

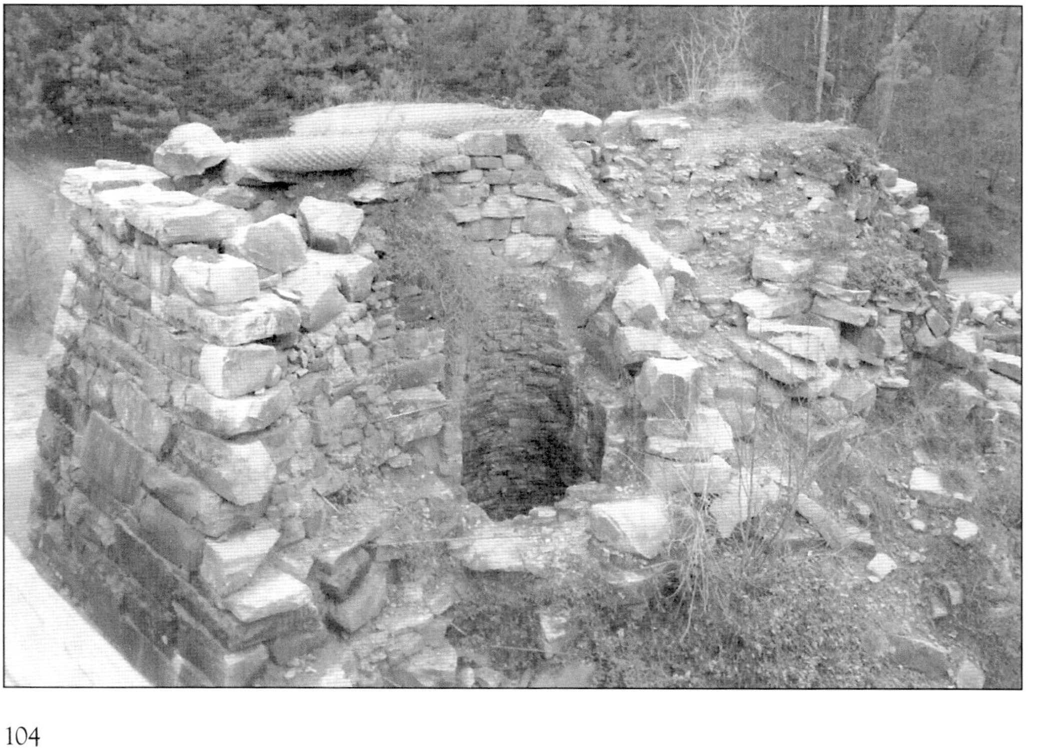

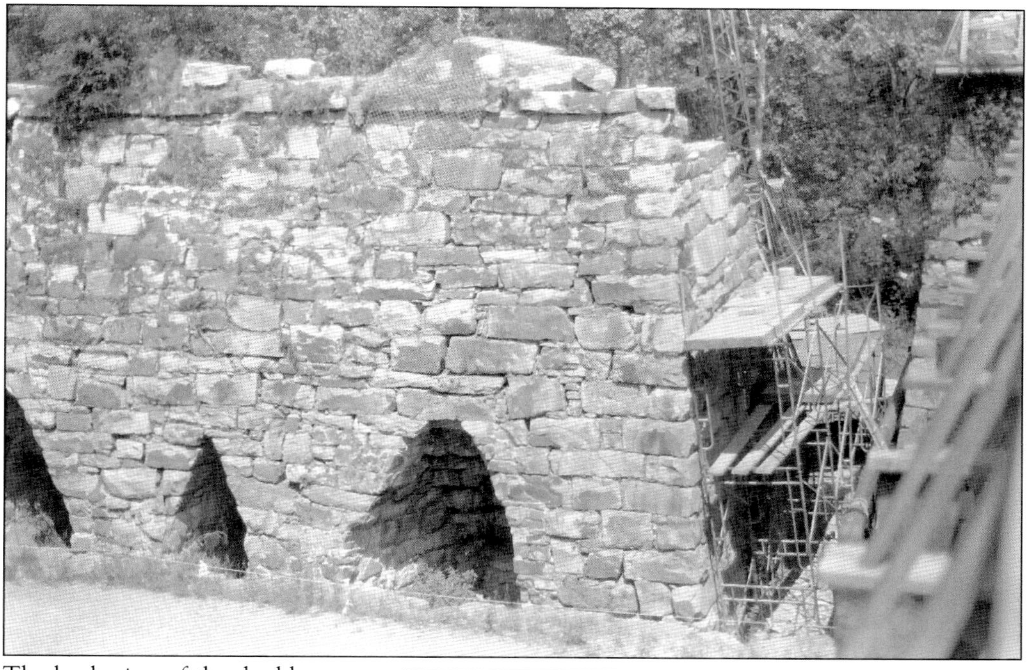

The back view of the double furnaces shows them largely in place from their Civil War appearances. Only the front walls fell in the 1930s, and the inner furnace portions, including an air passageway, remained stable. Here workers are repairing the fallen portions in 1985.

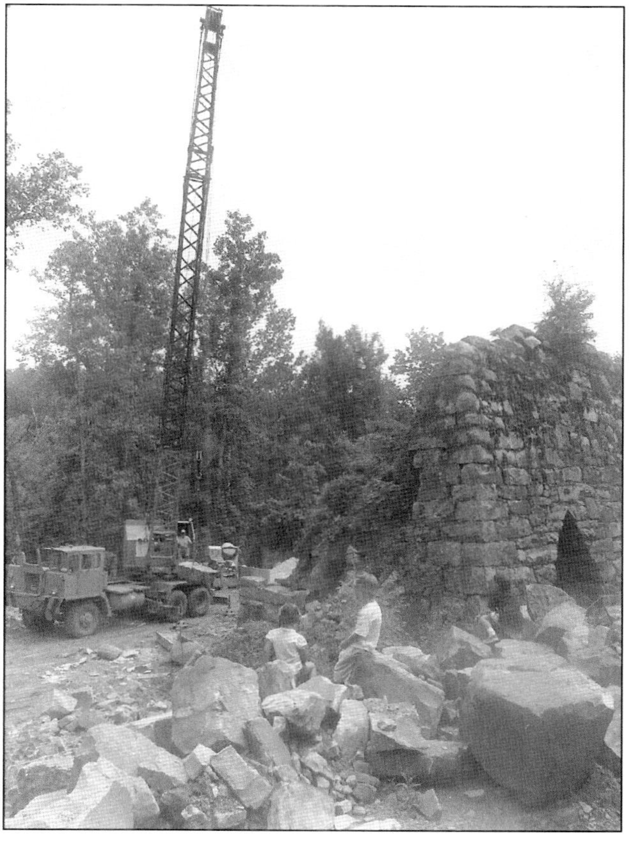

Children watch from a distance as a huge crane prepares to lift a heavy stone in the reconstruction of Furnaces Nos. 2 and 3 at Tannehill, which date to 1862. The restoration work ranks among the largest projects of its kind in American industry. When completed, the Tannehill Furnaces took on their 1865 appearance.

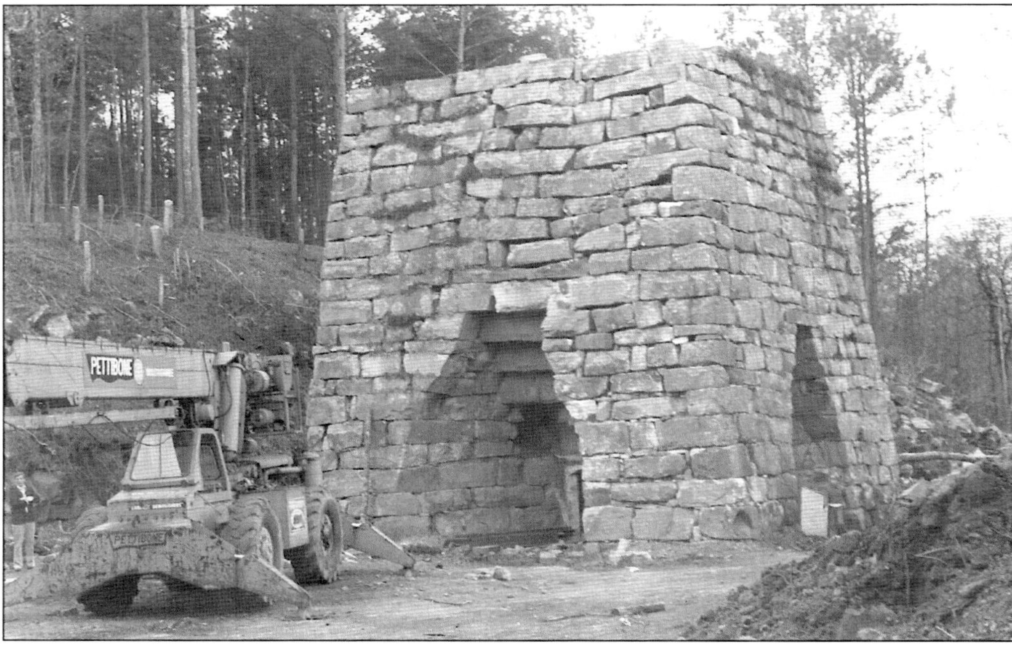

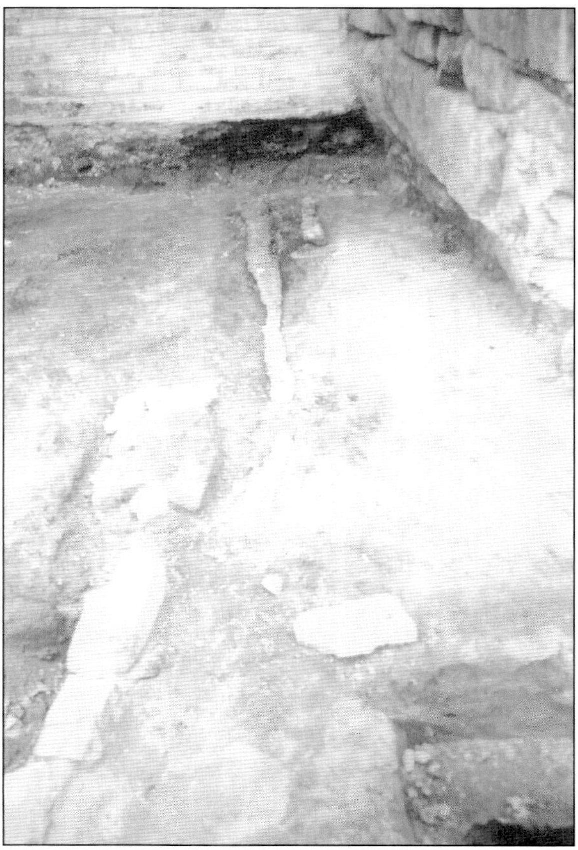

This picture of Furnace No. 1 in 1975 shows progress being made on its restoration. The southeast corner was badly damaged by vandals who used dynamite to discourage interest in a public park at the site in 1968. Work crews are getting ready to build back the draft stack and the charging bridge on the hillside.

The last run of iron from the Tannehill Furnaces may have been uncovered by archaeologists at Furnace No. 2 in 1994. The iron "runner," the last portion of a furnace's molten charge, extends into the casting floor. This charge probably took place in the latter days of March 1865—just before the arrival of Union troops.

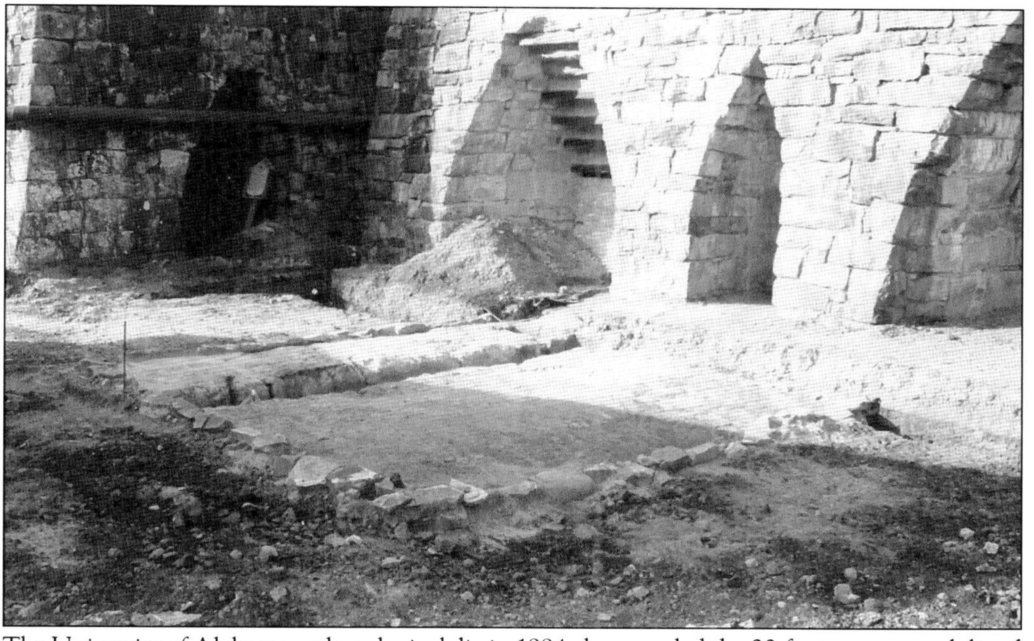

The University of Alabama archaeological dig in 1994 also revealed the 20-foot, square, rock-lined casting floor for Furnaces Nos. 2 and 3 from which thousands of pig iron ingots were cast from 1862 to 1865. The beds held casting sand into which ingots were cast as well as specialized shapes using cast boxes, some of which have been recovered. The area was once covered by a huge shed.

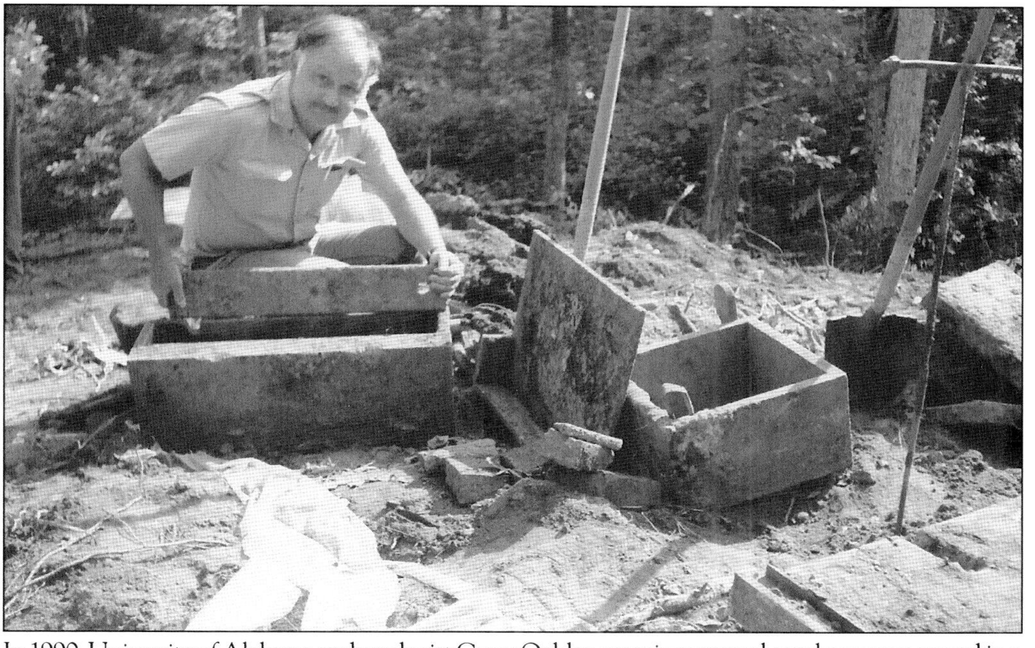

In 1990, University of Alabama archaeologist Carey Oakley examines several cast boxes uncovered in a pile of artifacts, probably placed there either by scrap hunters or by company employees of Republic Iron & Steel Corporation in the 1940s. Republic scrapped tons of artifacts in its Thomas Works but overlooked this collection. More than a foot of dirt hid them from public view for almost 50 years.

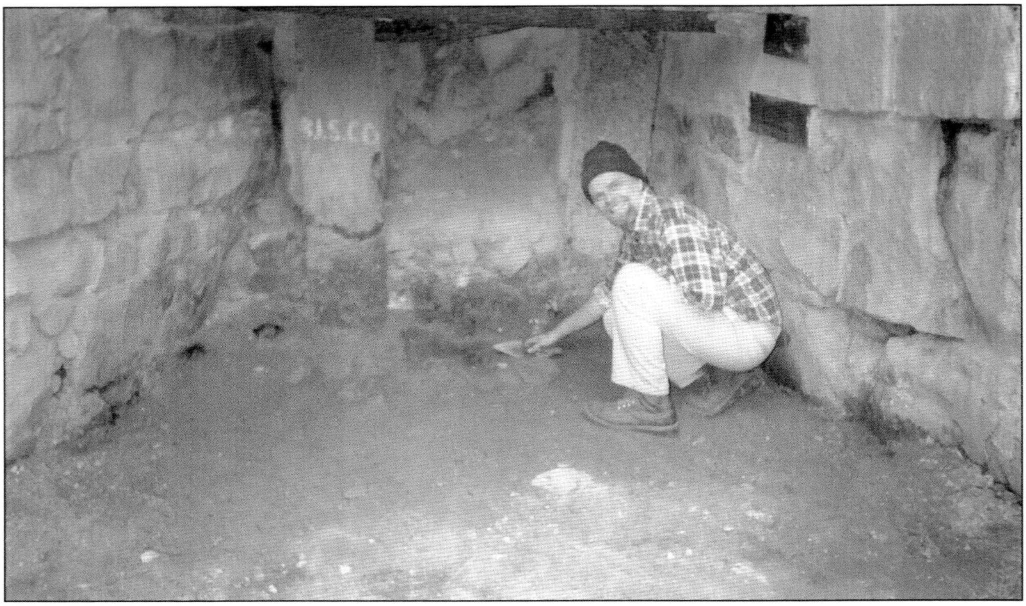

In 1994, University of Alabama–Birmingham archaeologist Jack Bergstresser brought to light several new discoveries, including a rare clay plug used to hold back molten metal and the presence of mantel rings inside the heat chamber. These rings were an innovation in Southern iron manufacture in the Civil War period. Before its removal, the site had been protected by fallen stone since the 1930s.

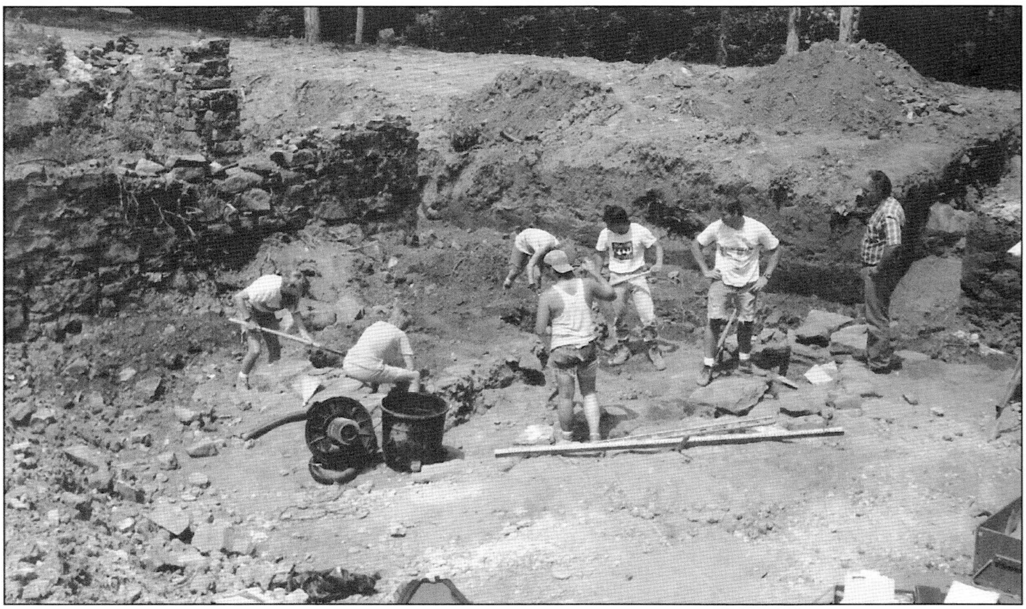

University of Alabama archaeology students gained valuable experience as part of the team of investigators in 1994 that found the charging beds in front of Furnaces Nos. 2 and 3. The area had been cleared of fallen stone for the first time since the 1930s during restoration activity, which took place in 1985–1987. The investigation also revealed that the double furnaces had been built over an area partially filled with slag, leaving the ground unstable and contributing to their fall.

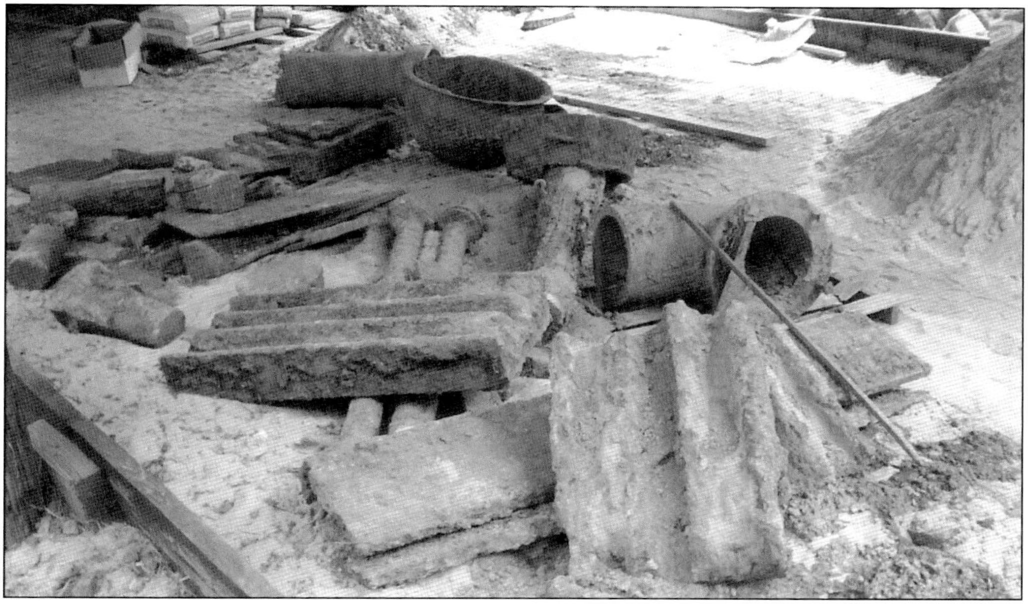

These artifacts were among 1,074 individual pieces recovered during the University of Alabama's 1990 archaeological investigation at the Tannehill Furnace site. They include pipes, kettles, rail, sledge blocks, forge anvils, and various cast iron pieces. In addition, park employees came across more than 50 more during the reconstruction of Furnaces Nos. 2 and 3. Only a fraction of the total artifact collection is on display at the Iron & Steel Museum.

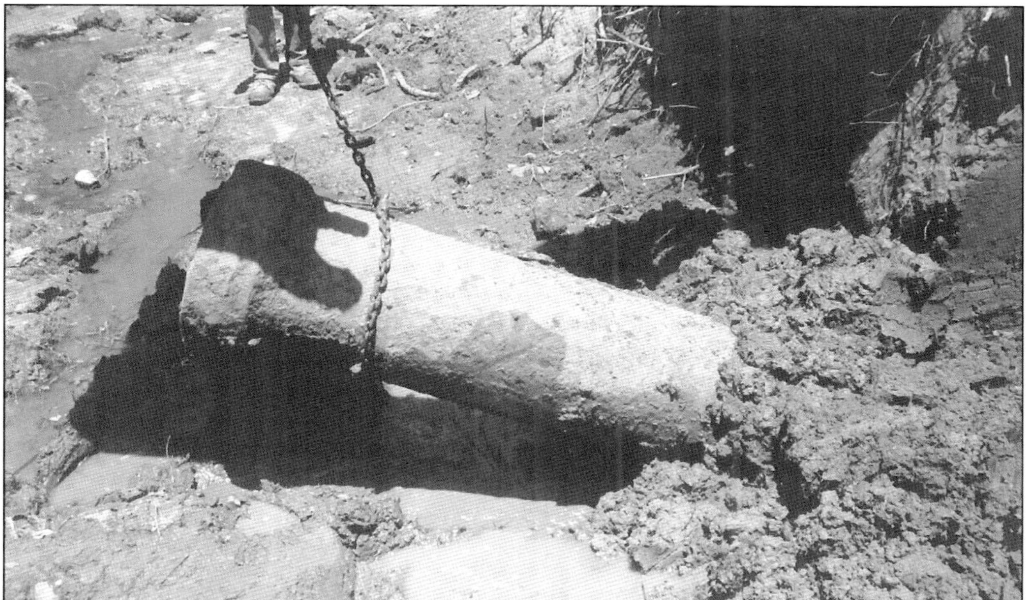

Among other items recovered was this complete eight-foot section of 16-inch-diameter cast iron pipe, probably used in the transmission of air blast. Various elbow pipes and connectors, along with three steam pipes measuring seven feet long by 2.5 inches in diameter, were also found near the blower house and its adjacent water canal.

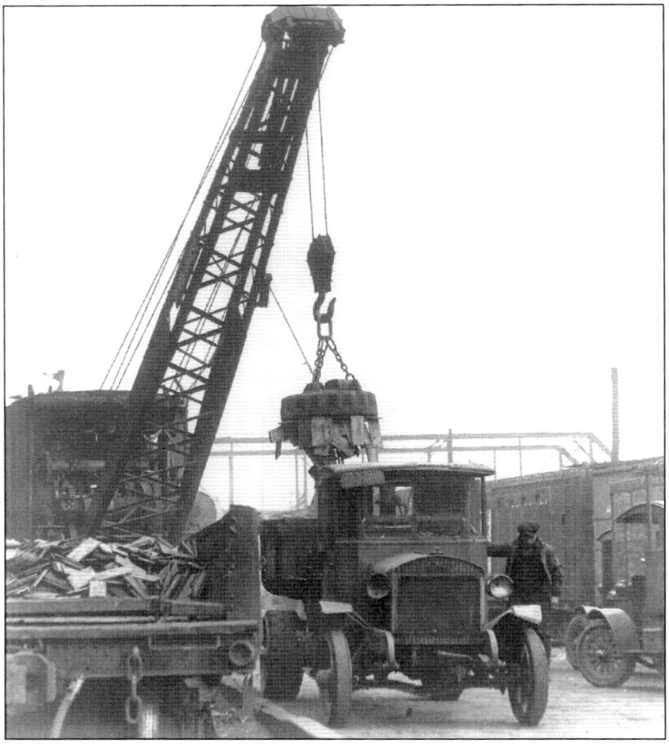

Scrap recycling during World War I began taking its toll on machinery and blast pipe left from the Civil War period at the Tannehill Ironworks. It continued through World War II. It is feared that many rare machines and tools used in the iron-making process here were lost in scrap drives. (Institute of Scrap Recycling, Inc., Washington, DC.)

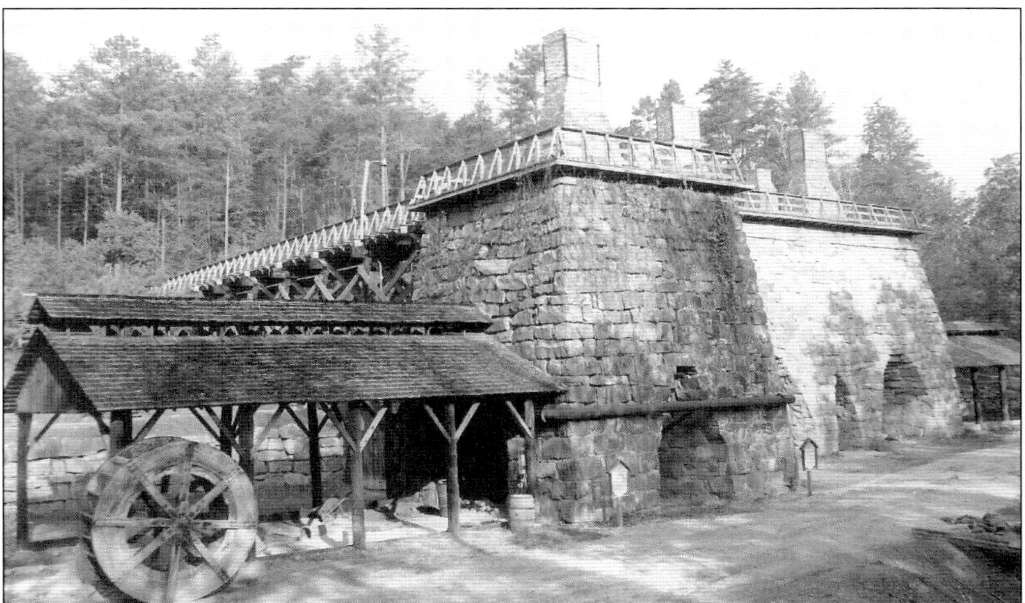

This view of the furnace yard around 1988 shows the furnace restoration work completed, but the bridge house construction had not yet begun. That work, along with putting a roof on the charging bridge, was completed under a $250,000 grant from the Save America's Treasures program of the US Department of the Interior in 2005. Additional improvements were made in 2011.

This rock-lined canal running from the blower house to Roupes Creek was dug in the 1840s to support the original bloomery; however, it was also used for Furnace No. 1. Likely dug by slave labor, the water channel carried water from the blower house back to the creek. It probably began to fill with dirt shortly after a flood in 1866 and was not uncovered again until 1990.

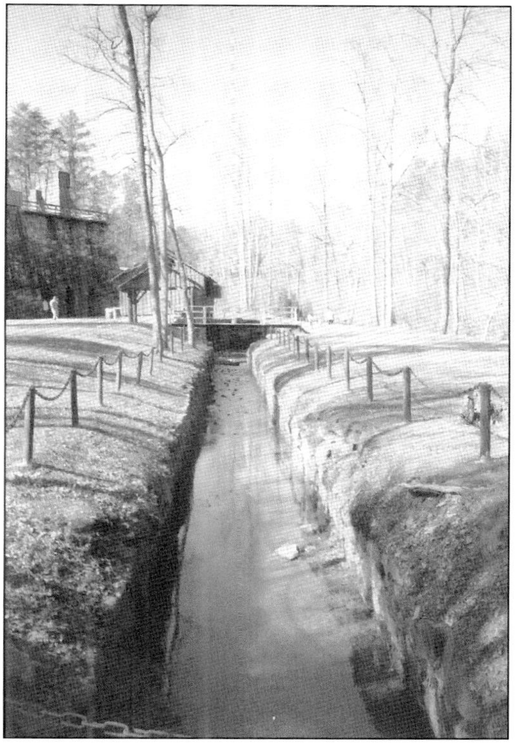

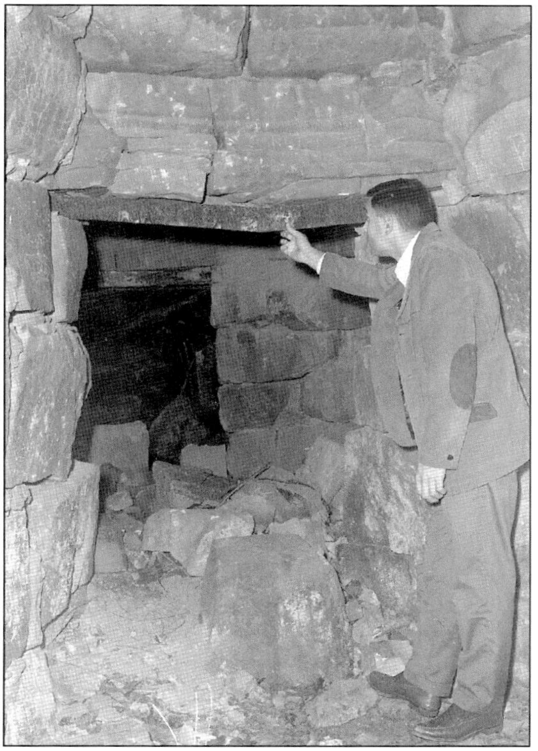

An engineer checks a bent mantle piece over the crucible in the 1862 furnace. The damage was probably caused by the 1930 collapse of the front exterior. Tons of concrete added in the restoration process, however, have stabilized the structure for years to come.

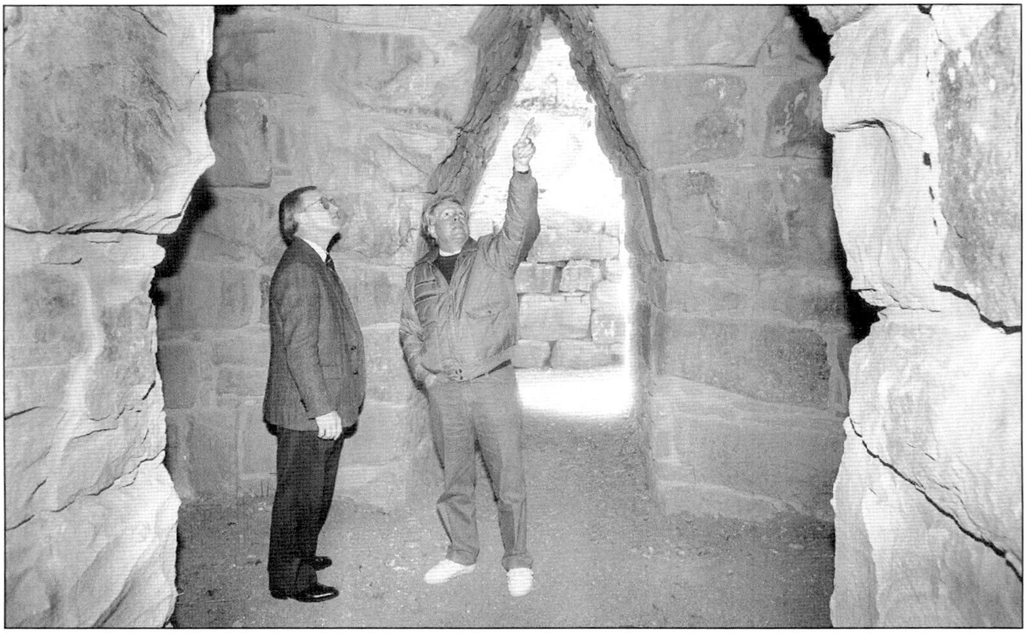

In this 1990 photograph, taken in the air passageway inside the double furnaces, park commissioner Jim Bennett points to structural details with Red Fifield, the general manager of US Steel's Fairfield Works. The area not only was used for stoking the furnace charge but also as an updraft for furnace gases. The US Steel Foundation is a major donor to the restoration work. (US Steel.)

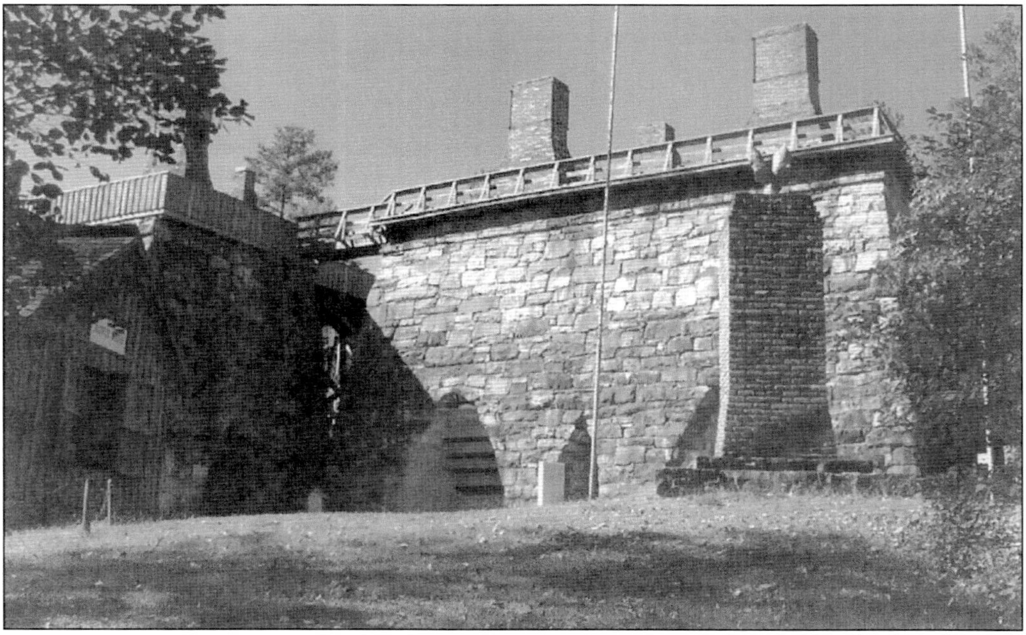

The Tannehill Ironworks in this photograph are as they appeared in 1995. Phase II of the restoration plan in 2005 built back the bridge houses and covered charging bridge in a major renovation project. Future phases include reconstruction of the front and south cast sheds, the raw materials warehouse, and the water flume.

Six

Refiring of Furnace No. 1

The suggestion that the older furnace at Tannehill be restored for an actual run of iron was first made by state geologist Walter B. Jones in 1952. While the idea was seriously considered, it would not become a reality until Tannehill Park planners proposed it be made a part of the American Bicentennial Celebration.

To prepare for the 1976 event, the park set about restoring Furnace No. 1 to operating condition, including the relining of the bosh with fire brick and building back the draft stack, which had apparently been removed by early-day house builders seeking free bricks. The charging bridge was also rebuilt, as well as the cast shed. To pay for the extensive restoration, funds were contributed by the State of Alabama, the US Department of the Interior, the American Revolution Bicentennial Administration, the Alabama Historical Commission, and Birmingham locals of the United Steelworkers Union.

US Steel's Fairfield Works contributed 50 tons of iron ore, 20 tons of limestone, and 150 tons of coke. The American Cast Iron Pipe Company donated an industrial blower, and Abex Corporation contributed safety equipment for more than 50 volunteer workers from area industries. Ray L. Farabee, a professor emeritus in chemical and metallurgical engineering from the University of Alabama, served as furnace master.

Stroup's furnace roared back to life on September 16, and three days later, it yielded 2.5 tons of iron, the first it had produced in 111 years. The Smithsonian Institution said the refiring marked the first time in US history that an iron furnace, abandoned for over a century, had ever been put back into blast.

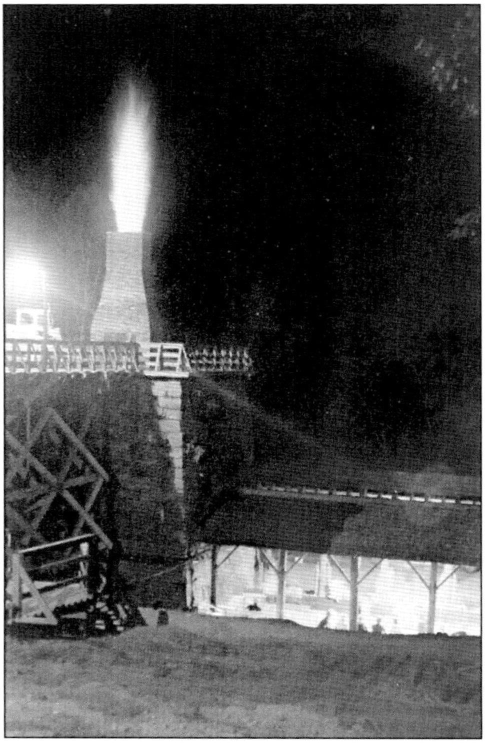

Fire shoots out of the top of Furnace No. 1 on September 19, 1976, as part of Alabama's participation in the American Bicentennial Celebration. By the end of the day, about 2.5 tons of pig iron were recovered, most of which went into replica cannons and souvenir ingots and bookends. Extensive restoration had been done on the furnace built by Moses Stroup in 1859. Over 50 volunteer workers from iron and steel companies in the Birmingham district spent two weeks getting the restored furnace ready for the historic run of iron. (John Farmer, Birmingham, Alabama.)

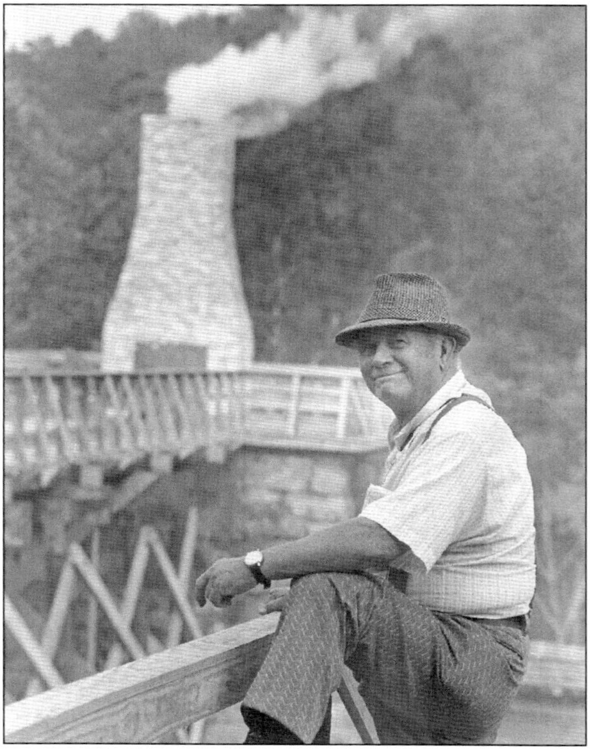

The furnace master for the American Bicentennial run was Ray Farabee. Professor Farabee, who died in 1994, recommended the Tannehill Furnaces to be nominated as a national historic landmark for the American Society for Metals. The status of a national historic landmark was given the year after his passing. (*Tuscaloosa News*.)

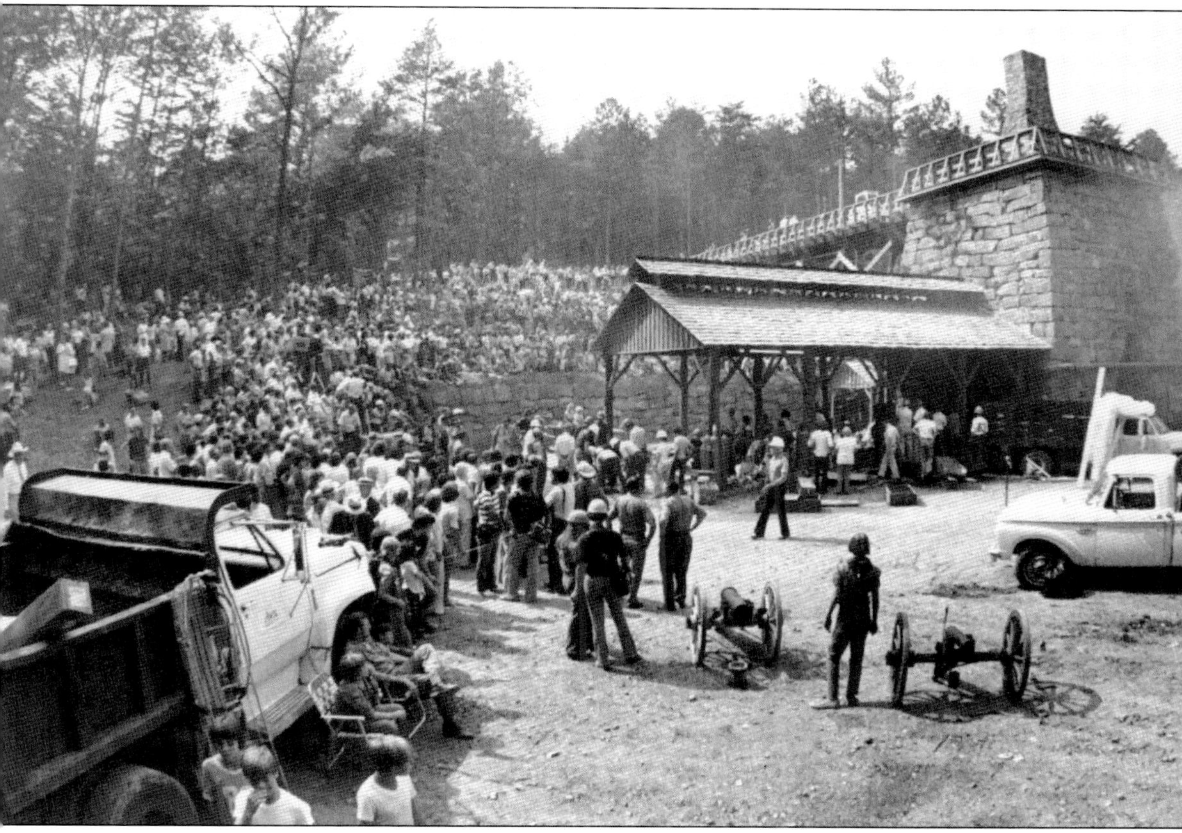

Some 15,000 people attended the recharging ceremonies at Tannehill State Park on September 19, 1976. Unfortunately, many others were turned away. During the refire, some 5,200 pounds of iron ore, donated by US Steel, were charged into the works. The event attracted national attention. (*Birmingham Post-Herald*.)

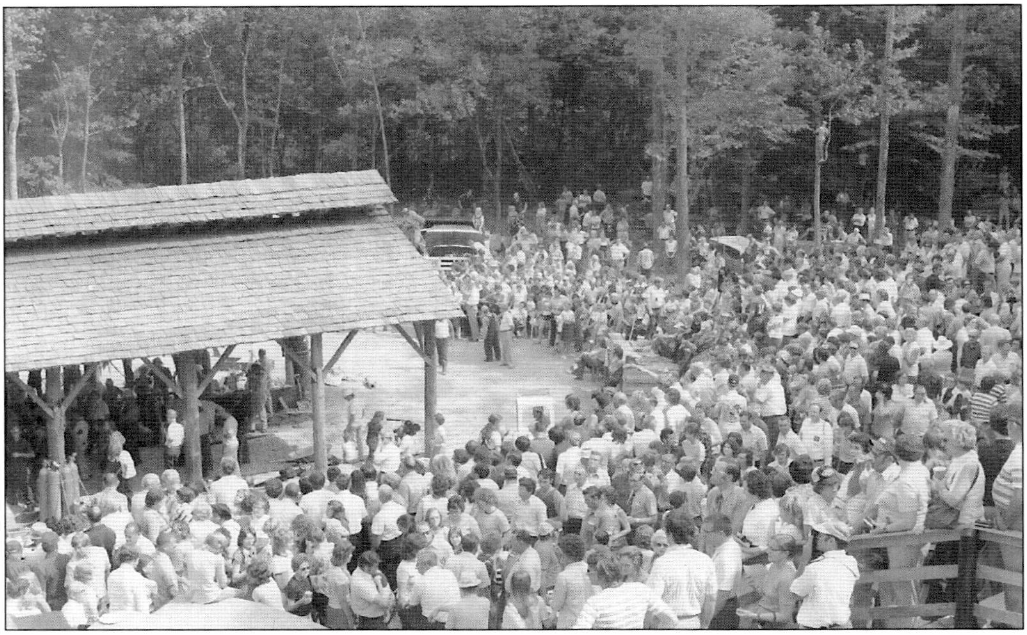

Shown here is part of the huge crowd that turned out for the Bicentennial refire tapping ceremony, which marked the first time in 111 years iron had been produced here. Federal troops put the works out of commission on March 31, 1865, as part of Wilson's Raid into the state to destroy Alabama's iron industry and its support of Confederate military needs. (*Birmingham Post-Herald*.)

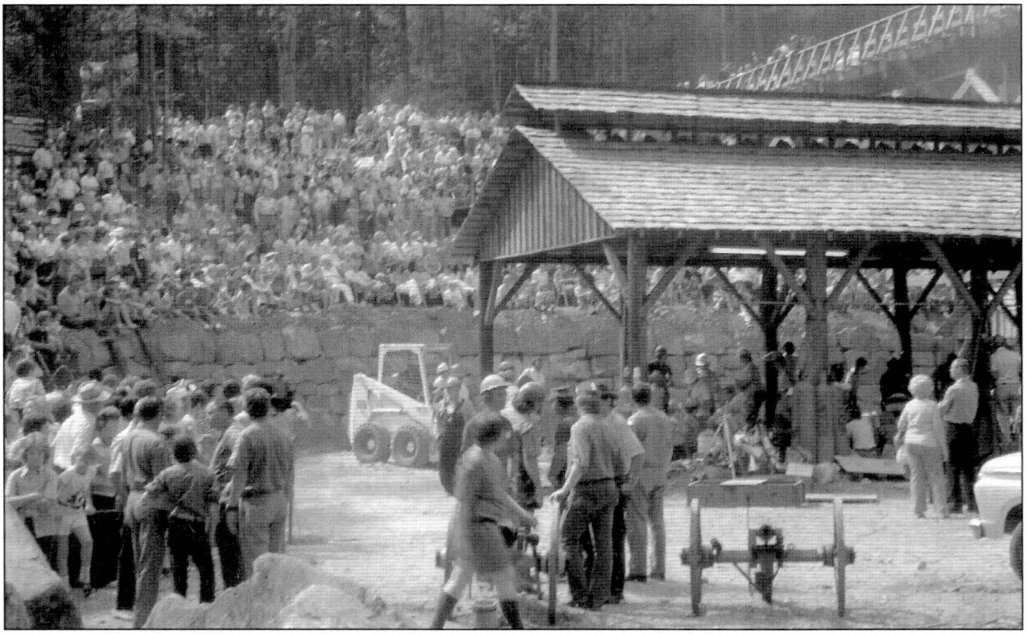

The overflow crowd waited for several hours as officials spoke of the old ways of iron making, and workers scurried to complete the six-hour reduction cycle, which produced several tons of pig iron. Blast furnaces in the 1860s were tapped every six hours and ran around the clock for months at a time.

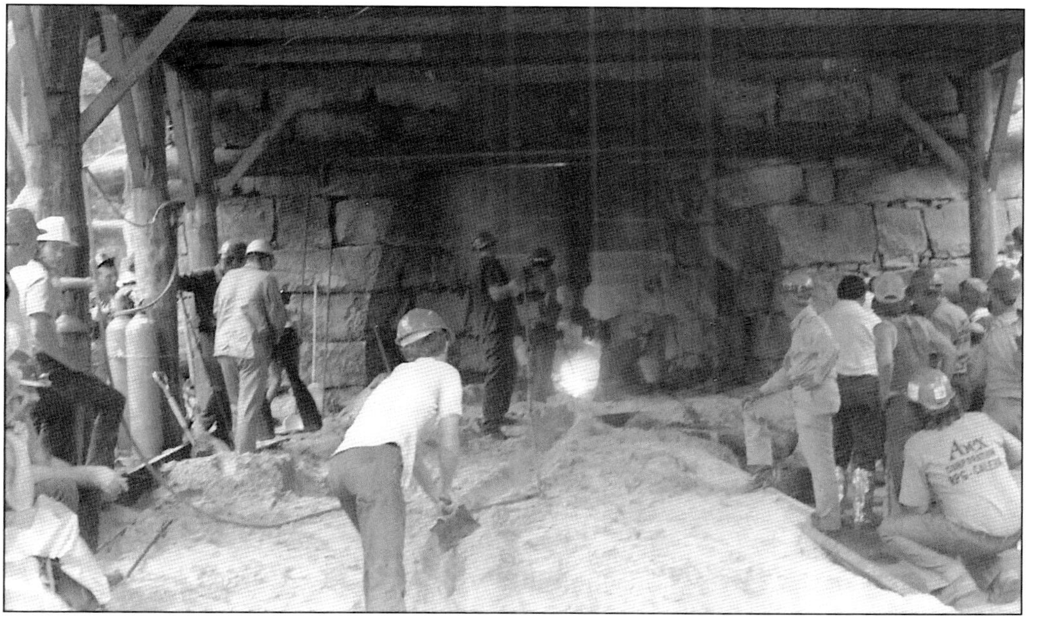

Here, some of the 50 iron and steel workers who volunteered to conduct the recharging of the old furnace prepare the sand beds for the first run of iron at Tannehill since 1865. The old furnace moaned and groaned as it came to life once again, sounds that had not been heard in more than a century of inactivity.

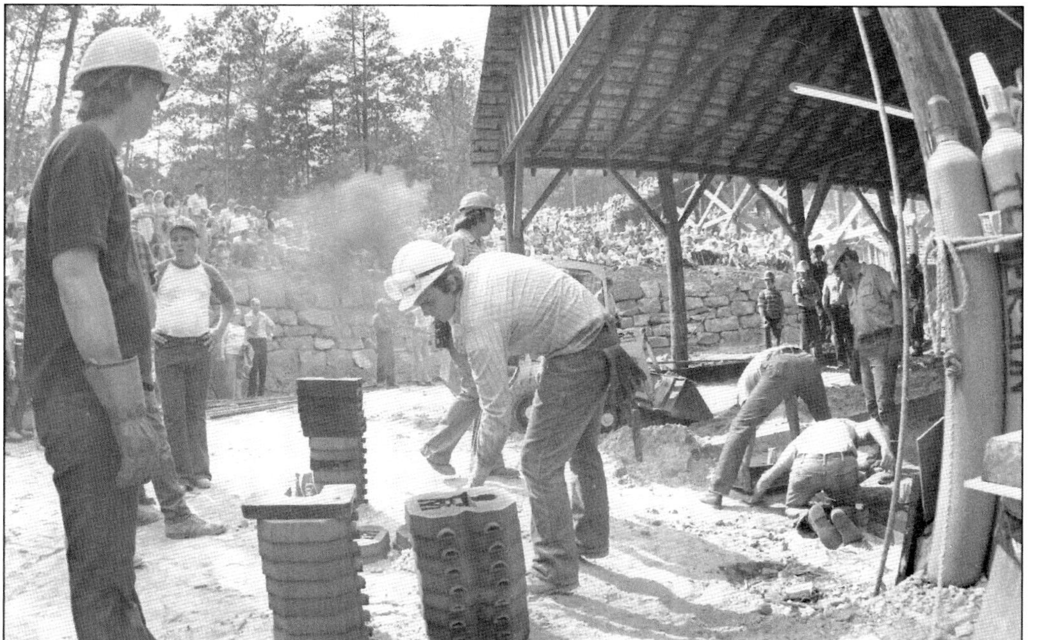

Foundry workers prepare for casting operations as the hot metal begins to flow from Furnace No. 1. Special mould boxes had been prepared for cannons, firedogs, and bookends. As modern ironworkers tried their hand at the work of their forefathers, it was fast and furious from the start.

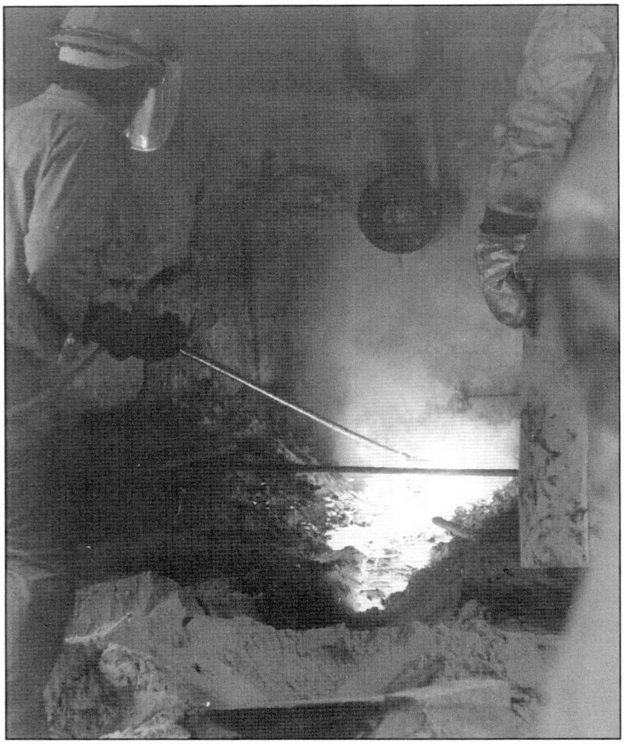

When the plug was pulled, 2.5 tons of molten iron came gushing out of the bottom of the furnace running into long runners. A newspaper photographer, who could feel the heat from 3,100-degree temperature, caught this historic moment. (*Birmingham Post-Herald.*)

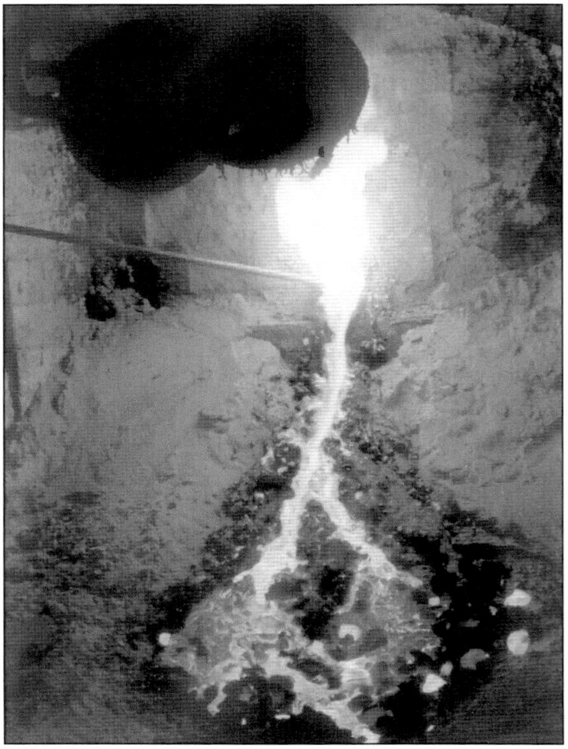

White hot metal flows from Furnace No. 1 during the Bicentennial run on September 19, 1976, as Alabama pays tribute to the nation's 200th birthday. It was the first time since the Civil War that iron flowed from the base of the old Stroup's furnace, to the delight of thousands of on-lookers.

Foundry workers for Stubbs Foundry in Helena and Abex Railroad Products Corporation in Calera cast several replica 12-pounder Napoleons and mountain howitzers from the run. One of each are currently exhibited on park grounds near the country store. Another Napoleon is used in Civil War reenactment ceremonies by Fowler's Alabama Battery. The men pictured are, from left to right, Sid Morgan, an Abex safety engineer, and Clyde Long, a project engineer. (*Birmingham News.*)

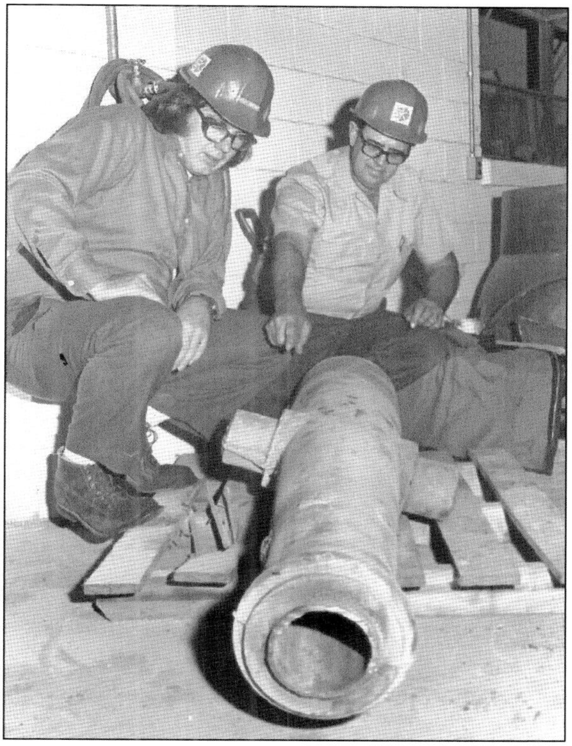

This mountain howitzer and 12-pounder Napoleon cannon were cast from the bicentennial run of Tannehill Furnace No. 1. They are now on display near the country store at Tannehill Ironworks Historical State Park in the Civil War Memorial. The Tannehill Ironworks was added to the National Register of Historic Places in 1972 and was placed on the Civil War Trust's National Discovery Trail in 1996.

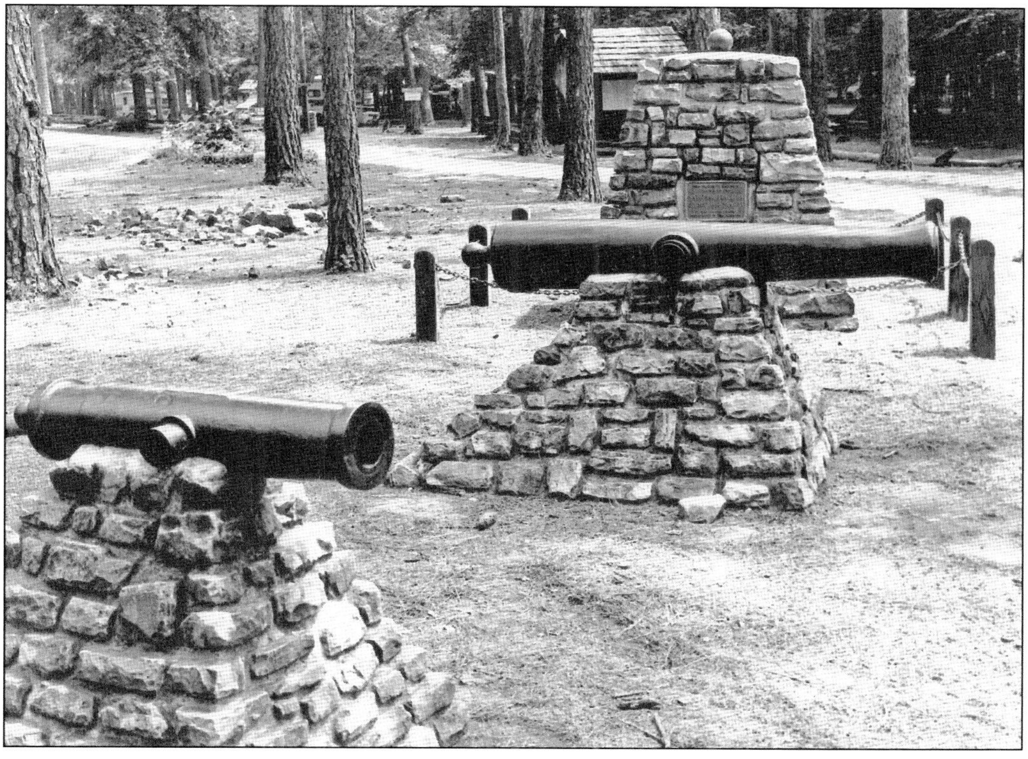

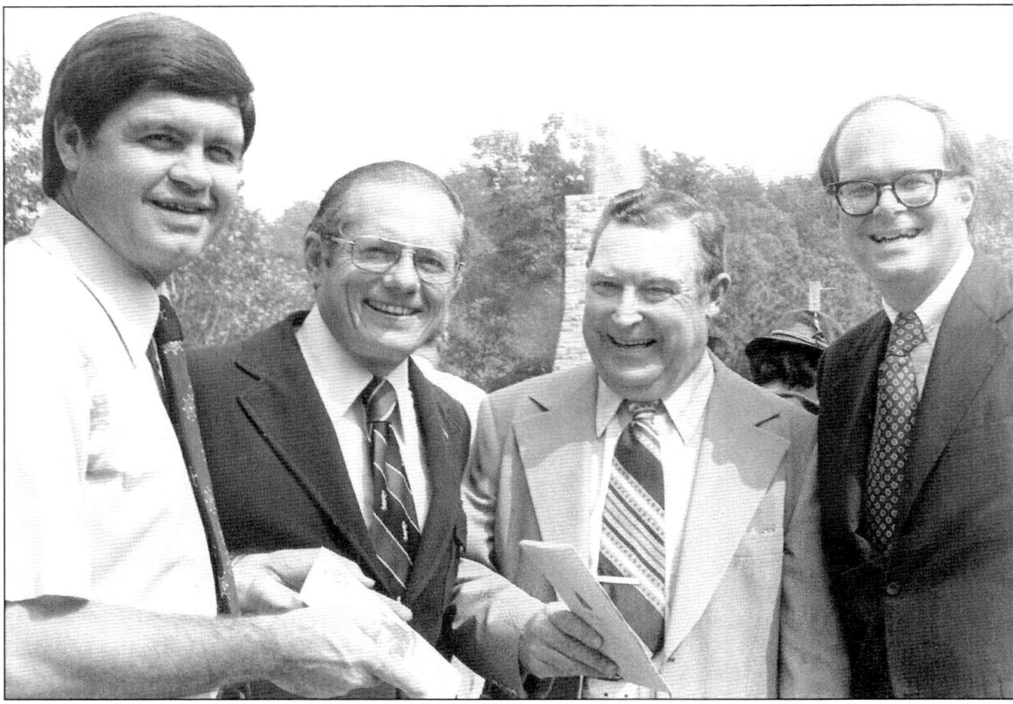

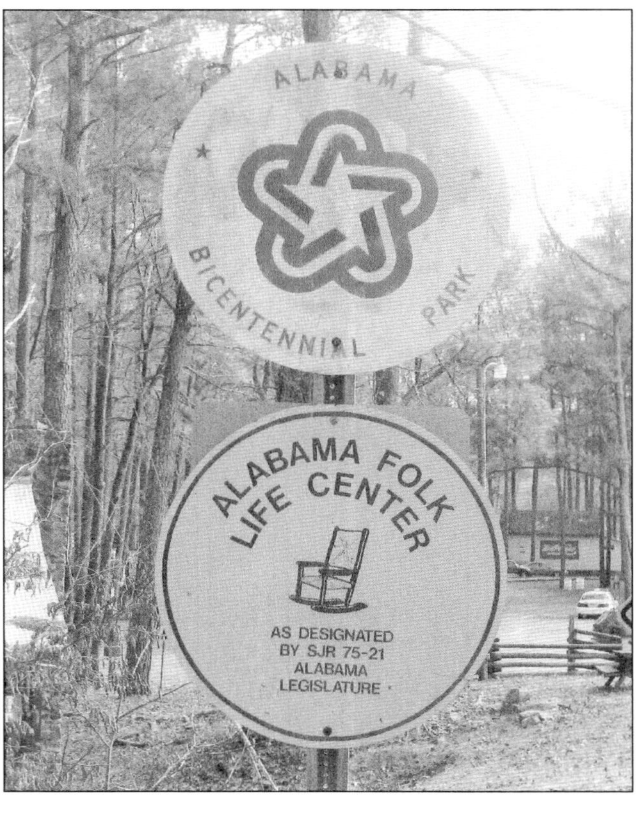

Among the platform speakers for the Bicentennial event in 1976 were, from left to right, Lt. Gov. Jere Beasley, US representative Walter Flowers, Gen. Taylor Hardin, and Douglas P. Wheeler, deputy assistant secretary of the US Department of the Interior. Master of ceremonies for the event was Leland Childs, a former state senator and local radio personality. (*Tuscaloosa News*.)

Tannehill State Park was designated a Alabama Bicentennial Park by the Alabama Bicentennial Commission in 1976 in commemoration of putting Stroup's 1859 iron furnace back into blast as part of the state's contribution to the nationwide observance. The Alabama Legislature also designated the park as "Alabama's official folk life center" in 1975. (Stacey Green.)

Seven

Tannehill Park Today

Now at 2,063 acres in three counties, Tannehill Ironworks Historical State Park has grown into one of Alabama's largest tourist attractions. Each year, over 400,000 people visit the park created by the Alabama Legislature in 1969 as a memorial to the state's iron industry.

It has become a national model on how an industrial site, abandoned for over a century, can be converted into a major tourist attraction.

With the restored ironworks as its centerpiece, the park includes 45 buildings, many of 19th-century origin, moved to the site to replace those burned by Federal forces during the Civil War. The attractions feature one of the South's largest collections of pioneer log homes, the John Wesley Hall Gristmill, the May Plantation Cotton Gin House, and the John Scott Young Country Store. The Pioneer Farm includes the park's blacksmith shop and the oldest building in the park, a milk storage building, dating to 1822.

The Iron & Steel Museum of Alabama contains thousands of artifacts from the state's iron and steel industry and one of the oldest steam engines in America. Some 33 major events are held annually at the park, including a Civil War battle reenactment noting the Federal attack on March 31, 1865, when three companies of the 8th Iowa Cavalry stormed the furnace yard. The old furnaces steadfastly remain.

Tannehill is a blast from the past.

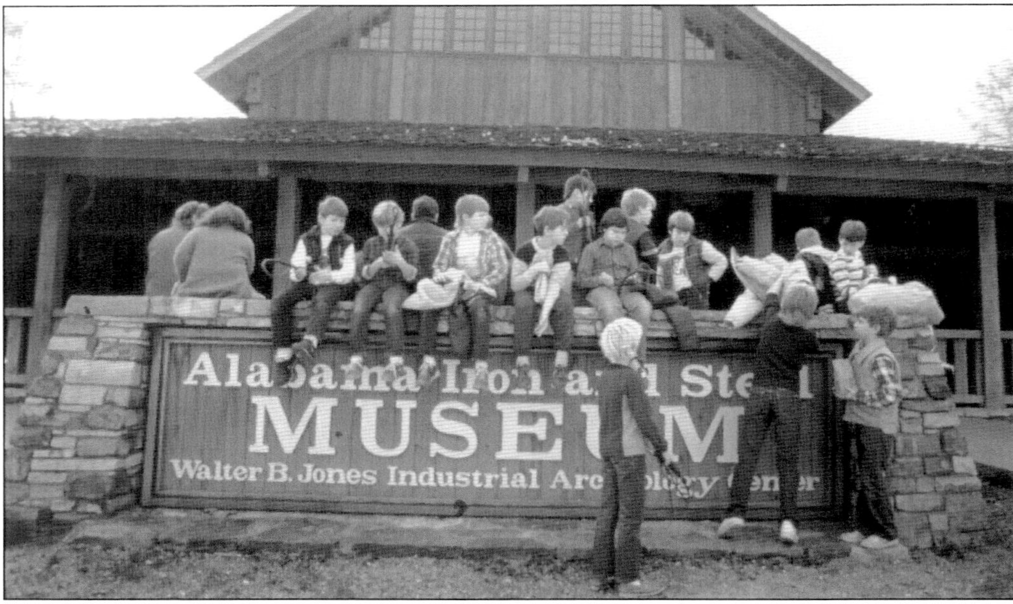

The Iron & Steel Museum of Alabama, built with a special appropriation from Alabama Legislature in 1981, is a Southeastern repository of iron-making technology from the 19th century. Its award-winning collections range from rare steam engines and Confederate artillery shells to an 1860s machine shop. It is also a key component of the Tannehill Learning Center, which annually attracts thousands of school children for out-of-class learning experiences. (*Birmingham News.*)

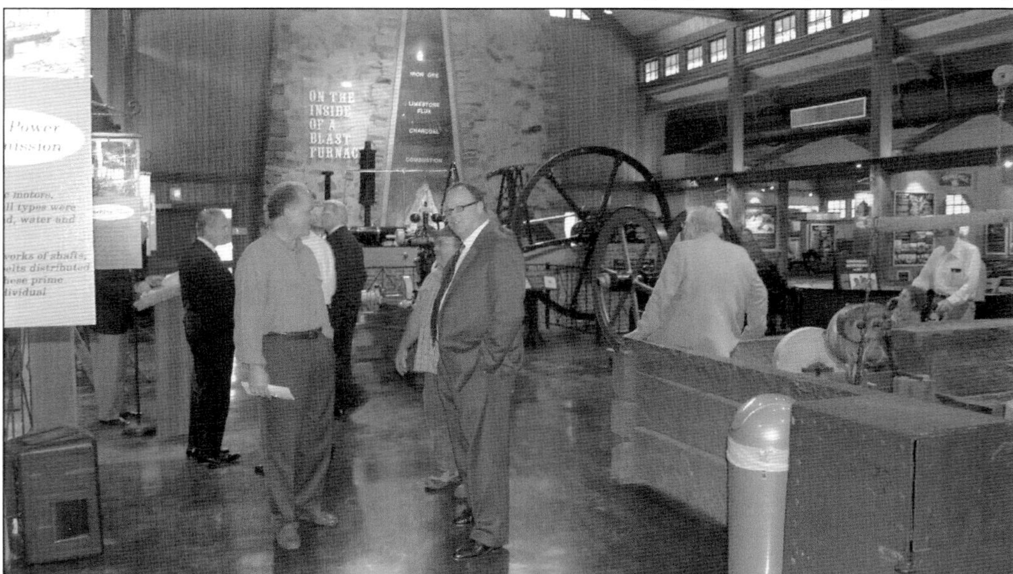

Undergoing major renovations in exhibits in 2005, the Iron & Steel Museum tells the story of the Alabama iron industry from the state's first blast furnace in 1815 through the present time. Birmingham's cast iron pipe industry makes 70 percent of the nation's water pipe, and the No. 8 Furnace at the Fairfield Works is the largest in the South. Former park manager Marty Everse (left) and current park manager Tom Land (right) are shown in the foreground.

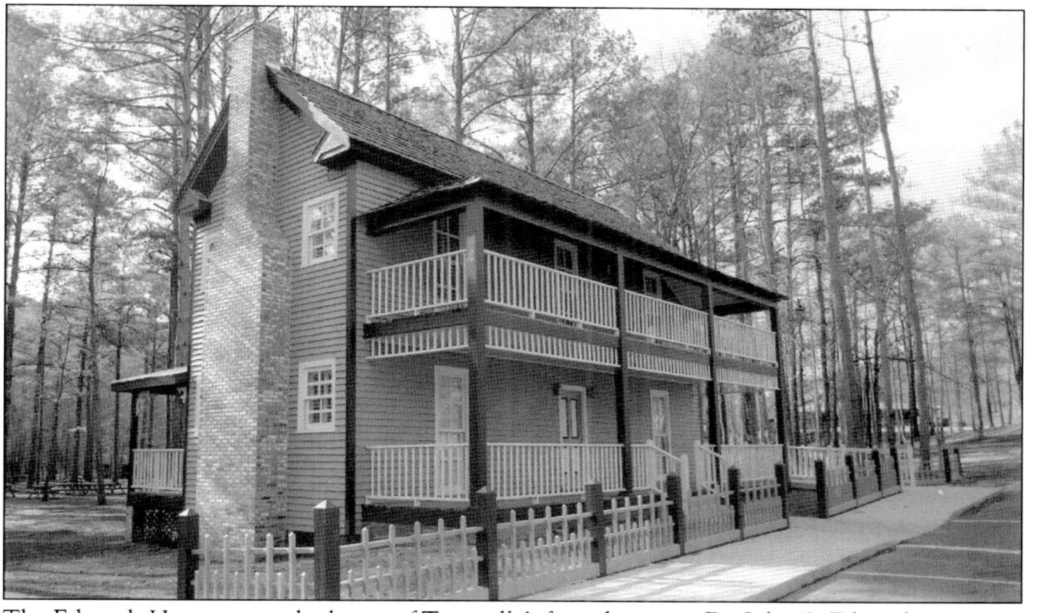

The Edwards House, once the home of Trussville's first physician, Dr. John S. Edwards, serves as the administrative building for Tannehill Ironworks Historical State Park. Built in 1878, it was moved from its original location on Chalkville Road in 1994 under a grant from the Alabama Power Foundation. The vernacular two-story farmhouse sat abandoned for years and was thought of locally as a haunted house. (*Tuscaloosa News.*)

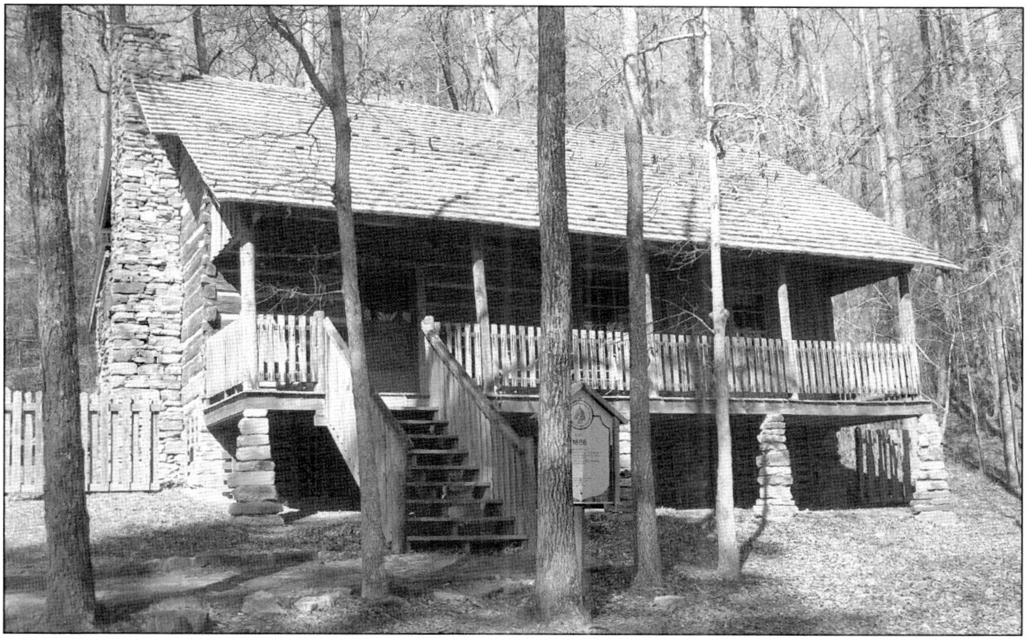

The Peel Cabin, built in 1890 in Shelby County, is one of a half-dozen original homes available for overnight stays. Tannehill Park hosts the largest collection of 19th-century log cabins in the South. While some are used for rentals, others are used for administrative purposes, education, and records storage.

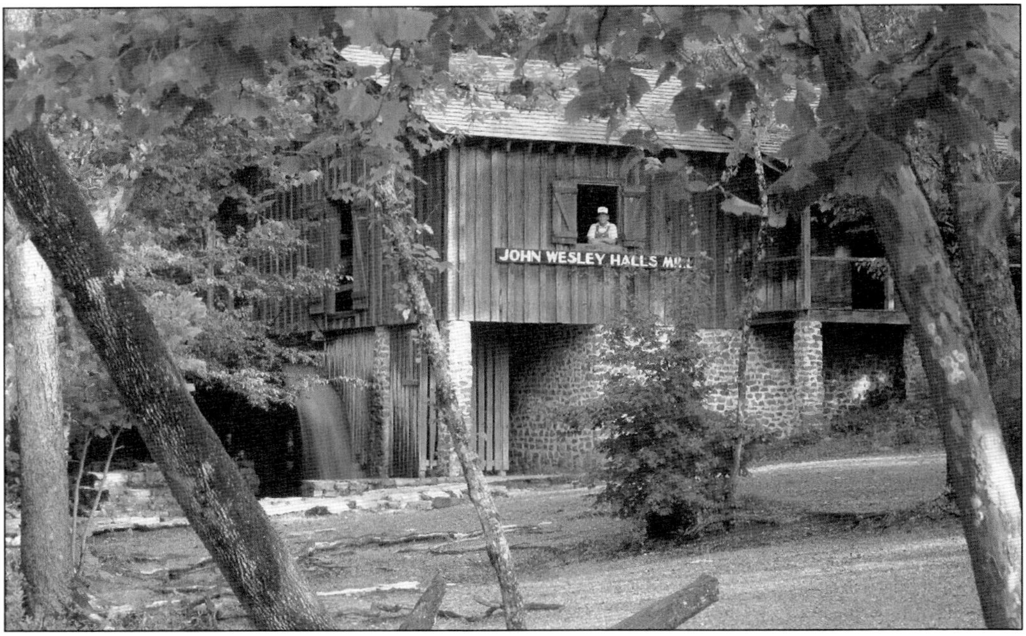

With a grant from the Linn-Henley Foundation, the Reconstruction-era gristmill operated by John Wesley Hall was rebuilt on its original site along Mill Creek in 1975. The older mill closed in 1931 after being damaged by a tornado. When this picture was taken in the 1980s, Sam Miller served as the miller. (Beth Maynor Young, Birmingham, Alabama.)

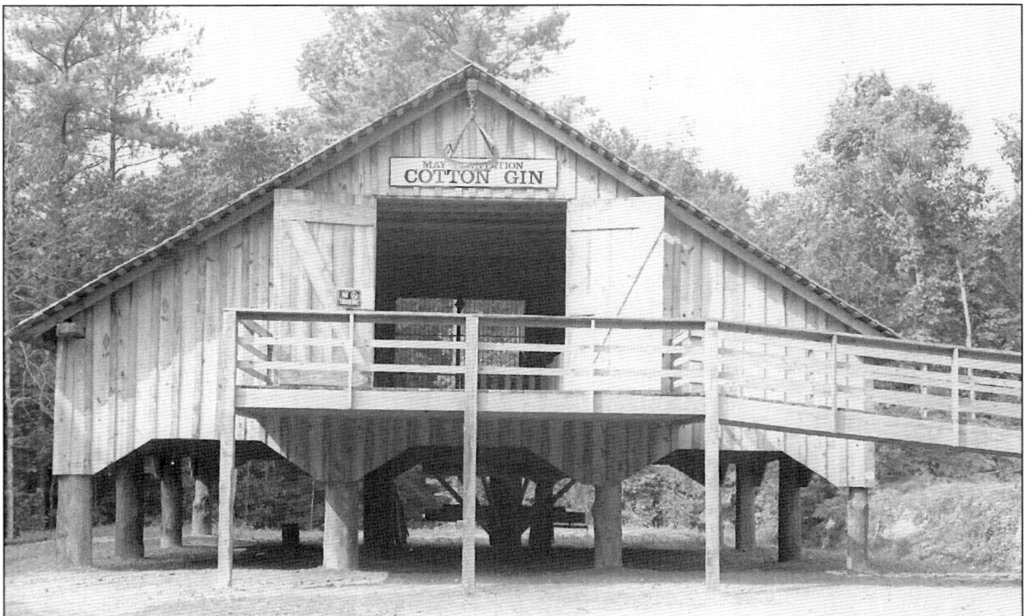

The May Plantation Cotton Gin House, built in 1858, was moved into the park from near Knoxville in Greene County in 1991. It houses cotton ginning equipment, including a rare 1881 Gullette gin head with feeder and condenser. The gin house utilized mules or horses to turn its machinery beneath the structure.

The John Scott Young Country Store, built near Brent on Highway 5 in 1905, closed its doors in 1941. Moved into the park and given new life as the park's country store in 1991, it serves the needs of campers seeking supplies and tourists looking for gift items. (*Tuscaloosa News.*)

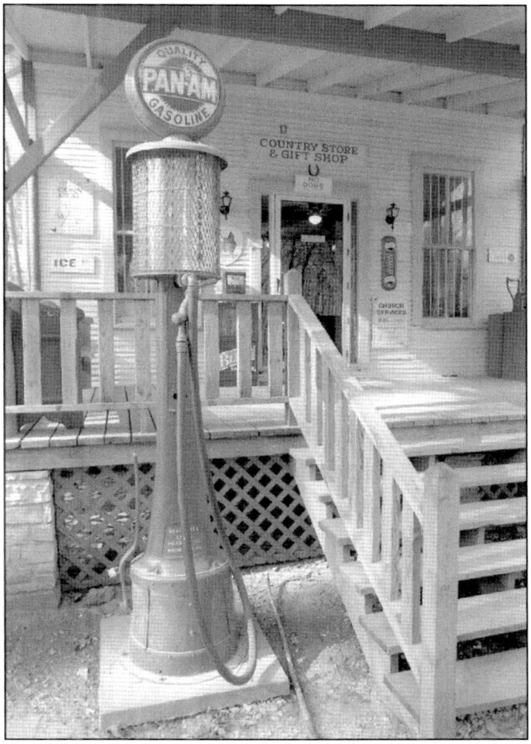

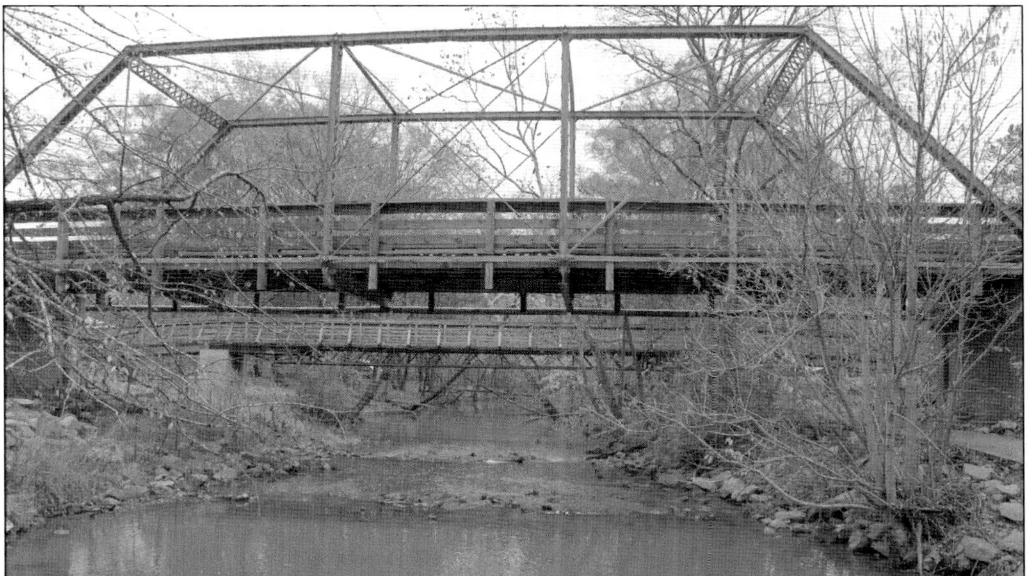

In 1975, Jefferson County moved the old Tapawingo Bridge (1902) into the park to span Mill Creek at the gristmill site. The bridge, over which thousands of Birmingham area residents crossed Turkey Creek in the early 1900s, is a classic example of an early iron truss. Tapawingo, located along the old Blountsville Road in Pinson, was a favorite turn-of-the-century swimming hole and resort. A second footbridge handles pedestrian traffic. (Jeff Newman photograph.)

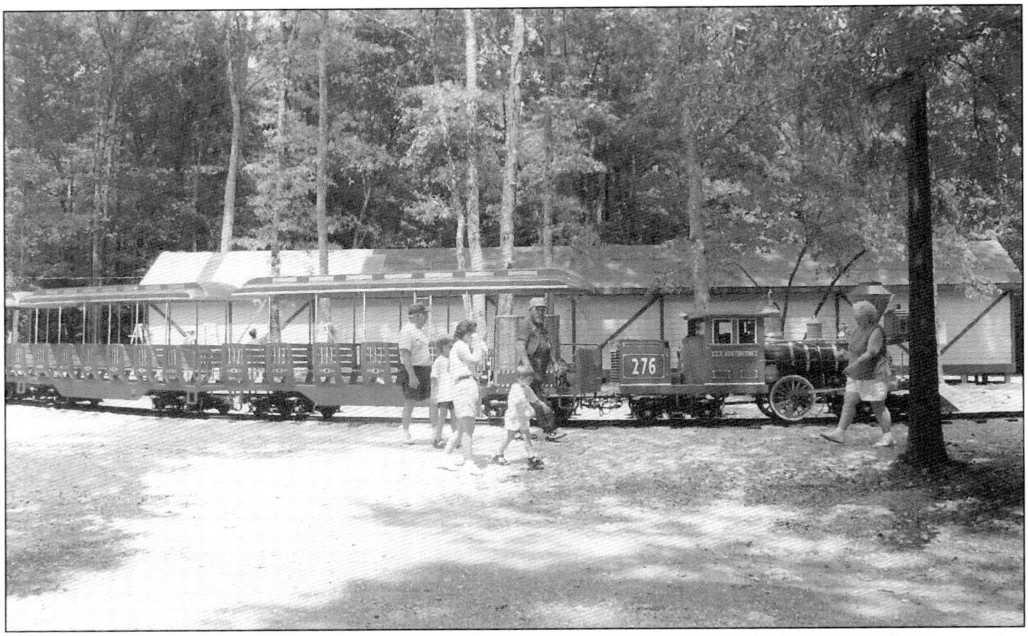

The Tannehill train, a three-car amusement park train, moves along a mile-long track from near the park restaurant to Farley Field, the site of monthly trade days. Added in 1997, it replaced an older train, "the Little Southern," purchased in 1978 from a city park in Guntersville. Hundreds of children and adults enjoy the train ride, which moves from Tuscaloosa County to Jefferson County.

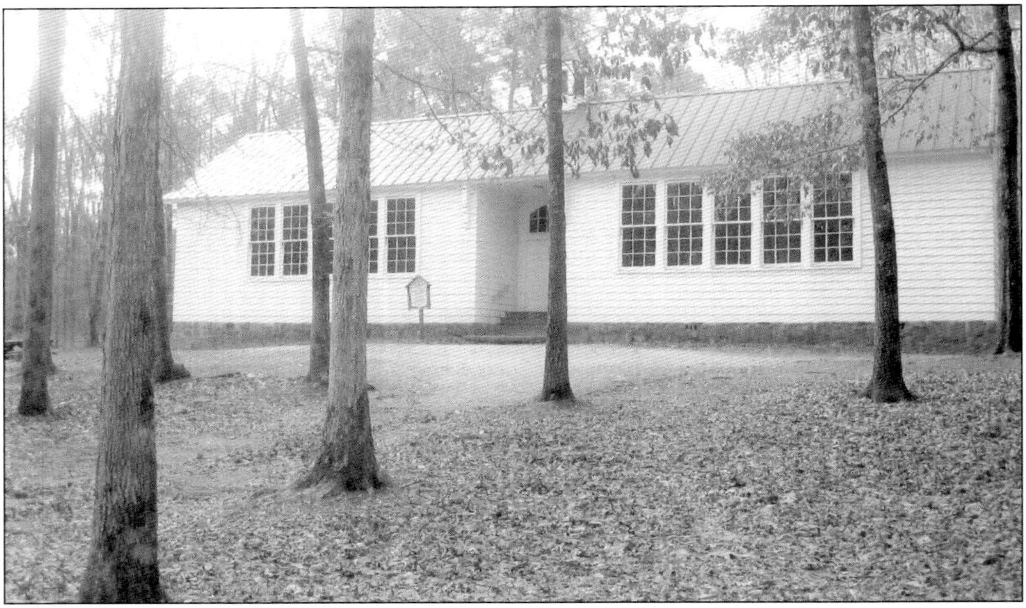

The Cane Creek School, built in Warrior in 1923, was added to the Tannehill Learning Center campus in 1983. It was successor to one of the first schools built in Jefferson County on land donated by William Thomas in 1815. The two other structures in the center are the Woodward Iron Company Post Office (1914) and the Fowler House, a log cabin built in 1860 near West Blocton. The log cabin was converted to a schoolhouse.

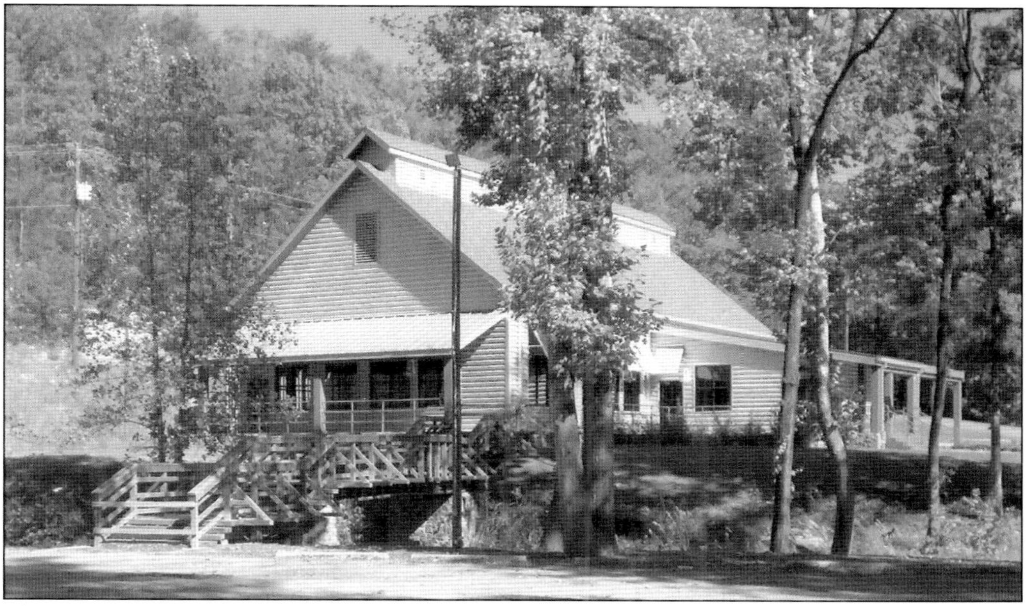

The new Furnace Master's Restaurant, built in the form of a forge, was added in 2005 to replace the original restaurant, lost to a fire in 1999. The 300-seat establishment is a "meat and three" country-cooking favorite for local residents and park visitors. It is managed by Betty Weeks, who was a partner, with her late husband Joe Weeks, of the first restaurant at the park. (Stacey Green.)

The old Kimbrell Methodist Church was restored on park grounds in 1972 after being moved from its original location about a mile away on Eastern Valley Road. Abandoned at the time, it dates to 1905. Kimbrell was the next community over to the east from Tannehill and was named for the same family that operated the Kimbrell Mill on Roupes Creek. Services are held at the church for campers and visitors on Sundays.

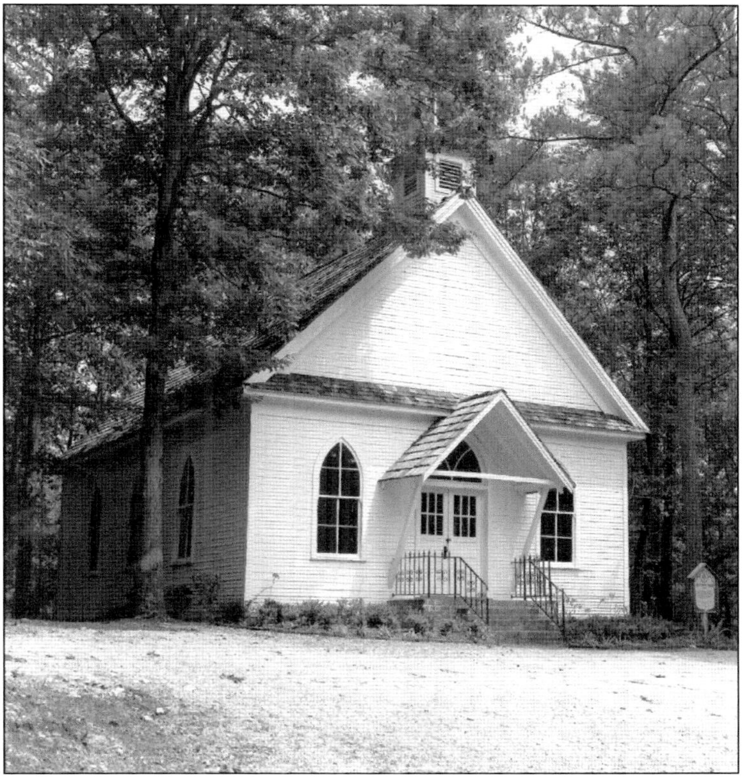